3 5/5

99¢

KILLER SALT

KILLER SALT

BY MARIETTA WHITTLESEY

INTRODUCTION BY MICHAEL LESSER, M.D.

BOLDER BOOKS

This book is for C.A.T.

I also wish to express my gratitude to
Dr. William H. Bailey of the Rockefeller University and
Dr. Godfrey Burns of St. Vincent's Hospital,
New York City.
My special thanks to Mr. Daniel Lockwood

Copyright © 1977 by Marietta Whittlesey.
Published by Bolder Books
a division of Hampstead Hall Press, Ltd.
38 East 57th Street
New York, N.Y. 10022

Library of Congress Catalog Card Number: 77-72258
ISBN: 0-918282-00-4

Book design: Esther Mitgang

First Printing

Distributed by Whirlwind Book Company, Inc.
PRINTED IN THE USA.

TABLE OF CONTENTS

INTRODUCTION

By Michael Lesser, M.D.

Marietta Whittlesey has performed a public service by calling attention to the great risk for our physical and mental health from common table salt. As Ms. Whittlesey documents in scholarly detail, many of us use far more salt than our body requires, and such common overuse of sodium chloride can contribute to a wide range of illness, from hypertension to depression to premenstrual distress. Even people still young enough to feel "o.k." may be insidiously endangering their future health without knowing it.

In this exciting, often personal detective story, Ms. Whittlesey not only demonstrates how salt can harm your body, but by demonstrating the relationship between improper nutrition and disease, and by cataloguing some of the dangerous substances which have crept into our regular diet, she provides vital information for all of us who want to safeguard our health.

Ancient physicians recognized the importance of proper diet in maintaining good health. Hippocrates, the Father of Medicine, told his students over twenty-five hundred years ago, "Let thy food be thy medicine." And the great Jewish physician of the Middle Ages, Moses Maimonides declared, "Anything that might be cured with food (diet), should not be treated by any other means." Despite these admonitions, modern medicine has become negligent in warning the public of the risk to their health from improper diet. People alarmed by the inexorably rising incidence of arteriosclerosis, diabetes, and cancer need look no further than our highly refined, frequently polluted and unbalanced modern diet for the probable explanation.

vii

Western physicians who should be leading the struggle for public nutritional education are too often sadly ignorant themselves, as they receive almost no training in nutrition in medical school. There is increasing evidence that everything from mental illness to cancer has some relation to faulty nutrition. Nevertheless, the priorities of modern medicine seem directed to finding relief for diseases once the disease already exists (and all too often, after it's too late) rather than providing the nutritional understanding to help their patients stay healthy in the first place. The medical student is taught how to perform surgery and prescribe drugs. But many of these drugs are completely alien to the body's ecology and have terrible, toxic side effects. Indeed, the really good drug therapies are actually derivatives of old-fashioned natural herbal remedies (foxglove—digitalis, valium—valerian root, etc.)

But as Ms. Whittlesey shows us once again, "money is at the root of the evil." A drug company which develops a new drug can patent that drug and then "sell" it to prescribing physicians. It's impossible to pick up a medical journal without finding several pages of ads touting this or that new drug. The profits from these drug sales pay for further drug research. . . . Many of our nation's most prestigious researchers maintain their laboratories partly or totally from drug company grants.

Nutrients, on the other hand, are plentiful in nature, inexpensive and *not* patentable. So there's no profit incentive to educate the public about the value of good nutrition.

Big food processors must share the blame for the great public ignorance about proper nutrition and the dangers of salt. Processing food removes much, if not all, of the natural tastiness as well as the nutritional value of food. Consequently, the big food companies must add salt (and sugar) to give their product some semblance of a recognizable taste. Unfortunately, the more salt we use, the more we need. Salt is an addictive drug, a drug for which we develop a tolerance. That's why the more we get the more we want.

Salt addiction originates in infancy. Ms. Whittlesey points out that baby food itself is laced with salt . . . and *not* for the infant's benefit. The salt is put in to make the food tasty to Mom and Dad, whose own taste senses have been perverted by years of using food that already has salt in it, then adding more salt while cooking it, then dumping on even more salt at the table.

Evidence exists that this kind of diet, with its resultant deficiencies in minerals such as zinc, creates a dulled sense of taste, requiring us to coat our food with salt in order to achieve some kind of taste which we normally associate with eating. And in a study performed by the National Institute for Health, administering 100 mgm. per day of zinc *did* result in a significant improvement in the sense of taste among persons with decreased taste acuity (hypogeusia).

Quick foods, packaged foods, canned foods, many frozen foods are not healthy for us or for our kids who are bombarded with TV commercials urging them to eat as much of this junk as they can get away with. Read the labels! Highly refined cereals, for example, that are forty percent sugar, are dangerous. These "convenience" foods are only convenient to the food processors. Refined foods are refined so they can be stored for great lengths of time without spoiling. They don't spoil because there is nothing left in them to spoil.

Even rats won't touch refined flour. Preservatives are added to kill bacteria, but what do they do to other forms of life, like human-kind? Poison is poison, after all. . . . In a carefully documented study, two-thirds of the laboratory rats fed a typical American commercial "enriched" bread were dead within ninety days . . . and the survivors were severely stunted.

Unfortunately, many nutritional scientists, who should be the watch-dogs of our dietary habits, also enjoy financial support from the big food companies. In return, these compromised scientists write hypocritical "scare" articles warning the public about "health food nuts" who dare to treat disease with nutrients, further misleading the public as to where the real danger lies.

This excellent book takes a giant step towards rectifying this sad state of affairs. Though I have been a practitioner of nutritional medicine for many years, I found much new and helpful information in this book, and urge thoughtful physicians and the public at large, to read this informative, exciting and possibly life-saving document.

Michael Lesser, M.D.
San Francisco, 1977

Michael Lesser is a graduate of Cornell University Medical College and received his training in psychiatry at the Albert Einstein Medical Center in New York City. In addition to his private practice and his work at the Napa State Mental Hospital, Dr. Lesser is President of the Orthomolecular Medical Society and President of the California Orthomolecular Medical Society.

PREFACE

The facts you are about to read in this book are shocking. If thoughtfully considered, reading KILLER SALT could prove one of the most important steps you will ever take in improving your health and sense of well-being. When we are dealing with our health we are dealing with the most precious of all personal possessions and so nothing in KILLER SALT is left to guesswork. As startling as some of the information may be, it is all based on known and proven scientific evidence.

Bluntly stated, nearly every one of us is endangering his or her health through the uninformed use of salt. Salt is considered by most a harmless condiment but it is in fact a powerful chemical agent, capable of doing the human body tremendous harm and *often causing death!* If this strikes you as an astonishing assertion, as it would have struck me before my careful investigations, I can only urge you to read what follows carefully and to weigh the medical and scientific evidence presented in KILLER SALT. If you do, I know you will come to see that salt is one of the truly unknown perils in our daily lives, that our consumption of salt is based on mis-education and *addiction* rather than on need, and that our susceptibility to salt is an important ingredient in many multi-million dollar businesses. Without doubt, salt causes disease—hypertension, kidney failure, anxiety, obesity.

I have no salt shaker in my home and buy no table salt. I don't use salt for cooking, preserving, and I never add it to my food. My body gets all of the sodium it can possibly use just from the foods I eat. In the back of KILLER SALT I provide some of my favorite salt-free recipes. I have also included a fairly comprehensive table listing how much sodium is contained in most of the foods we eat.

The chemical makeup of salt is sodium and chlorine. The part of this that your body uses is, of course, the sodium. Much of the illness so many of us experience—or will experience—comes from excessive sodium in our diets. Throughout this book, I will explain how sodium affects health and how different common illnesses are directly related to our sodium consumption.

But I think it would be best from the outset to be very clear about one crucial fact: everyone needs *some* sodium. I never use any salt on my food but that is because I know I am getting sufficient sodium from my regular diet. No one should ever under any circumstances attempt to put themselves on a one hundred percent salt-free diet. In fact, even before you drastically alter your salt consumption it would be, I think, a good idea to consult with a physician. Remember: Without *some* salt, the body's vital fluid balance will be critically upset and serious disorders can arise.

How much sodium do you need to keep your body running well? How much salt do you need daily to maintain vital functions? If you're an adult, your body runs best with about three ounces of salt in all of its fluids and bones. However, this salt is generally retained by your body and doesn't need frequent replenishment. Today, researchers and doctors are saying that we can be at our best with about one half gram of salt daily—in other words, about 1/8 of a teaspoon of salt. However, even this meager1/8 of a teaspoon will prove unnecessary if you are eating normally because your food will supply all of the salt you need.

If you are pregnant you will want to take particular care in making any sudden changes in your diet. Some research seems to point to the fact that in pregnancy we need more than our usual quantity of salt. Here again, I urge you to check with a physician. Also, if you are a vegetarian I urge you to pay particular attention to the sodium-food chart at the book's end. Without eating meat, fish, and milk products, it takes more careful planning to make certain of getting enough sodium in your diet. I believe that even my vegetarian friends will thrive on a diet which admits no added salt, but some care is necessary.

With these precautions in mind, I want to close this preface with the promise that if you carefully consider the evidence of KILLER SALT you will never again look at a salt shaker in quite the same way. My knowledge of the devastating effects of salt has dramatically improved my everyday life and I know it can make a huge difference in your life too.

Marietta Whittlesey
New York, 1977

Marietta Whittlesey is a journalist, a science writer and a former research assistant at Rockefeller University and at the New York State Psychiatric Institute.

CHAPTER ONE

The Salt Epidemic

Today, millions upon millions of Americans are suffering from the effects of salt in their diets. This national epidemic includes people from all walks of life, old people, young people, children, and even infants. Of the countless who are suffering because of salt, only a few will be able to help themselves and reduce their salt consumption. Why? Because only a small percentage of those who suffer from salt abuse realize what a crucial—and deadly—role salt plays in their life. Most of us consider salt just a harmless ingredient, something to be used in cooking, something to be sprinkled over our food. The commonness of salt and its virtually universal use makes it seem that it couldn't possibly be a "killer." Yet it is this very commonness that makes salt so deadly in our lives. For, truly, it is everywhere. It is difficult for most of us to even imagine a household without salt in it. It is simply a "given" in our lives. How could it be that something so common as salt could be for many a deadly poison? And if what we are learning about salt is true—that it *is* a chemical whose devastating effects are literally *frightening* to realize—how is it that salt is so commonly used and available? In other words, why weren't we told?

I believe that it is the birthright of all humans to live freely and well, to live with true joy. Salt in millions of cases drains and often ruins our health. And so the question naturally arises; why are we allowed to buy salt if it is so dangerous? Why is it everywhere? Why don't our physicians tell us more about this deadly compound?

I believe that to a very large degree the medical pro-

1

fession has been deficient in warning us away from the potentially deadly misuse of salt. Yet KILLER SALT is about the salt plague; it is not a book "against doctors." My own bias, when I began investigating the role that salt in our diet plays in our daily health, was and is basically pro-science, pro-medicine. My training was a scientific one and my livelihood has been in the field of medical research. Still, time and again, I was brought face to face with a startling fact; millions of Americans are being poisoned, sometimes slowly and sometimes not at all slowly, and the great institutions upon which we have learned to rely—government, industry, and even medical science—have done very little to alert and protect us.

In this respect, I can only applaud the current movement toward medical independence for the layman. While this book is written out of great respect for medical science (and indeed I have used much of what our doctors have already proved about salt and its role in disease), it is also undeniably true that if our government and the majority of our doctors were protecting us from the terrible dangers of excess salt there would be, quite simply, no need for this book.

But the fact remains that there *is* a need for Americans—for people everywhere—to be alerted to the dangers of excess salt. Surely, if nutritionists, government agencies such as the Food and Drug Administration, and the official sources of medical information were truly dedicated to exposing the proven and undeniable hazards brought on by the excessive salt in our diets, the information presented in KILLER SALT would be commonplace and redundant. And one day, I feel, our government and our medical societies will take over the duty of educating the populace to the dangers of salt, as well as other hidden and thus far unexposed perils in our diet. Until then, however, it is the responsibility of those doctors and medical researchers who have pinpointed the incredible dangers of excess salt to make absolutely certain that this information is spread and that the knowledge of this danger is understood by all.

Socrates said, "One cannot get closer to the gods than to bring health to one's fellow men." For one who attempts to explain the perils of salt there does, or should, exist a sense of profound responsibility. I would be, however, less than honest unless I admitted that my interest in salt and its role in health did not begin with an abstract desire to contribute to the common good. In truth, my concern with salt began in an intensely personal way. I was suffering, to put it plainly. Suffering badly. Before I discovered the dangers of salt and began to eliminate it from my daily diet I fell prey to some of the most uncomfortable and bewildering side effects generated by this silent killer. I suffered from fits of depression, unbearable tension, and humiliating weight gain that, in a very real sense, were ruining my life. In the course of this book I will share my own personal experiences with salt, and also I will share how I went salt-free and the changes this has meant not only in my physical health but in my overall outlook on life.

Q. *Who Suffers From Salt?*

As I began to explore the disastrous impact salt was having on my own life I also became newly sensitive to how others around me unknowingly harmed themselves with salt. One close friend of mine suffered chronically high blood pressure. Other young women friends—even the most liberated among them—accepted repeated premenstrual agonies as just part of being female. As I investigated the deadly impact of salt I soon realized that I was tracking down a killer and I wasn't its only victim.

I was on the trail of a potentially destructive yet wholly common substance whose victims were everywhere. The more I learned about salt and its effects the more deeply convinced I became that not only does it represent a severe peril for those who actively suffer from its effects—and the active sufferers range in the millions—but salt also undermines the health and sense of well-being of tens of millions who do not even consider themselves as ailing.

I am fascinated by scientific detective work and I had the great advantage of studying with a medical researcher whose major area of study was stress diseases. Following his clues and the quickly accumulating yet largely unpublicized findings of the medical profession, I learned how and why these stress diseases are aggravated by an excessive salt intake. As I pursued my research I had an opportunity—if witnessing such grim sights can be decently called an opportunity—to explore the hypertension and kidney clinics of New York City. There I saw women and men suffering horribly from diseases which might have been averted entirely if they had not exacerbated their body weaknesses with a tragic and uninformed overuse of salt. Yet the countless numbers who are already at this moment paying so dearly for their overuse of salt are only partial evidence of salt's awesome destructiveness.

I think it is reasonably safe to say that for every person already suffering from the effects of excess salt—be it in the form of damaged kidneys, high blood pressure, or cyclic weight gain, there is another person who suffers at a level just below detection. Such a person might not think of himself as sick, or even unhealthy, but he nevertheless lives with daily malfunction which effectively robs his life of true joy and energy. To such people, feeling good, truly and deeply well, is an unrealistic, utopian vision, a kind of medical fairy tale. So used to feeling "pretty good" or maybe even "a little draggy," these marginal sufferers may not consider themselves salt casualties but, in so many cases, they are. While the sufferer from excess salt who manages to keep out of the hospital isn't as tragic as those whose salt intake leads to serious and often fatal ailments, the "everyday" sufferings of millions are nevertheless a terrible—and altogether needless!—outrage against the body. Like all of the living creatures of the earth, we were born to be healthy.

Q. *Can We Help Ourselves?*

Recent history has proved that if we expect the best of health we must participate more actively in the quest for a well body. We can no longer—if we ever could—depend on government, industry, or even the medical profession to insure for ourselves lives of energy and good health. We must protect ourselves by knowing more and more about our bodies and how the substances we consume affect us. Few, if any of us, are capable of being our own physicians in times of real emergency, but with increased knowledge and a little care we can greatly reduce the damage we ourselves inflict on our bodies. And while understanding the role of diet will not make anyone an instant doctor, it is undeniably true that understanding diet will very often mean that we will not even *need* to see our doctors for anything more than an annual check-up. By explaining the effect of salt and by attempting to answer some of the most crucial questions relating to salt, I hope to bring you through a process of exploration and understanding that I myself went through.

I am convinced that all of us can be healthier, more vibrant, and happier in our lives than we are today. I hope that KILLER SALT will be one more step—perhaps a giant one—in that direction.

CHAPTER TWO

What Is Salt?

Table salt as we know it is actually the crystal product of the union between chlorine, ordinarily a poisonous gas, and sodium, a reactive metal which bonds easily with other elements to make new compounds. (If you take any kind of household salt and heat it up in a frying pan you'll soon be able to detect the distinct, chemical odor of chlorine rising from it.) Table salt, that utterly common stuff so many of us shake unthinkingly onto our foods, generally contains about 40 percent sodium. The body *does* need a certain very small amount of sodium to operate well. But the consequences of dosing the body with too much sodium are painful and dangerous. Adding to the peril of salting ourselves into a state of critical sodium overload, we must also contend with the sodium we get without table salt. (As previously noted, in the back of KILLER SALT you will find tables showing how much sodium is contained in everyday foods.)

For now, however, it is important to stress that sodium consumption in all of its forms has the same potentially disastrous effect as sodium consumption through table salt. (Monosodium glutamate, which, incredibly I think, is still completely available throughout the United States, can be a major and crippling source of sodium.) When we begin to watch and restrict our salt intake we must carry our attention to all of the various sodium compounds.

Q. *Is Salt Poison?*

Salt itself is not a poison except when other physical conditions exist which cause an excess of salt to be retained. From that point, further salt in your diet can produce serious and even deadly complications. When I began first to cure my own ailments through salt restriction and then to study the whole complex question of salt and the body, I learned, much to my surprise, that there is absolutely no harm and no danger in never *adding* any salt to your foods. The Bedouins, who have lived for centuries in the heat of North Africa, do not use salt and have no word for it in their tribal language. Quite simply, the meat in their diet supplies all of the sodium their bodies need, even with the hot, often arduous lives many of them lead. Likewise, the Eskimos probably qualify as the most salt-free of all Americans. They do not use salt as a seasoning at all because they get all they need from their diet, with its supply of marine animals and fish. Interestingly, the Eskimos live surrounded by salt and that fact seems to make it less valuable and even less interesting to them. Later on I will discuss the odd history of salt and its symbolic importance to human beings, but for now let us note that for certain civilizations, even our own country during its formative years, salt was so rare a commodity as to be worshipped and salt was added to food as if it were the very elixir of life.

Q. Do We All Use Too Much Salt?

The blunt and shocking fact is this: the average American consumes fully ten times as much salt as the human body needs. This tremendous and *perilous* overload is directly due to overuse of table salt and, even more frighteningly, the enormous quantities of sodium compounds that are added to our foods. Much of this added sodium is, for all practical purposes, put in without our direct knowledge—as food preservatives or "flavor enhancers."

When I began to eliminate excess salt from my diet I

found myself literally poring over the labels of the canned and packaged foods I ate, searching for the truth of sodium content. Often the offending compounds were clearly stated (in fine print, of course) but often the possible existence of salt was obscured by the catch-all phrase "natural seasonings." For so many of us, food is salted during processing, during cooking, and then once again before eating. Many people add potentially destructive doses of salt to their food before even tasting it. As these unconscious salt addicts will often admit, this is "merely" a habit. But what a reckless habit it is.

Since the day I decided I no longer wanted to struggle through life with less than perfect health and less than great feelings about my body, I began observing how my friends and loved ones abused salt at the table. They had no idea what a powerful chemical was in their hands as they unthinkingly shook the salt onto their meals . . . and shook it and shook it and shook it. I'm quite certain that if you observe *your* friends and relatives with an eye toward their misuse of salt you will find the same thing.

Salt in itself is not a deadly compound. Salt is a vital compound, necessary for life. There *is* a crucial balance of salt to fluids within each of our bodies. This balance must be maintained at all times. If there is not enough salt in the system the body will self-correct this imbalance by conserving salt and losing water. The important thing is always the maintenance of the salt–fluid balance and it is one of the wonders of our bodies that they automatically seek this balance. If the body is dehydrated—from sweating, for example, or vomiting—the body automatically retains salt in order to store fluids and reestablish the salt–fluid balance. Lost fluids are, of course, crucial for the body to replace or else we would risk death by dehydration. The uncommonness of death through dehydration is testimony of the body's uncanny ability to sustain and regulate itself.

An interesting sidelight on the body's use of water and salt is the consumption of salt water. More than one shipwrecked soul has tried to quench a thirst through drinking sea water, with its high concentration of salt, and the result has been death. The fact is that we can, in an emergency, drink sea water but *only* if we can definitely drink fresh water within a short time. The salt concentration of sea water is intolerable to the body in its constant process of maintaining the salt–fluid balance, and the human organism reacts to sea water by vomiting and urination in an attempt to flush out the killing levels of salt. This self-purging leads only to further dehydration and eventual death.

Q. *Do We Need Extra Salt If We Are Hot Or Dehydrated?*

Dehydration is a condition that the body seeks to avoid and, naturally, we must exercise some care so as not to allow our supply of fluids to diminish greatly. This is fairly commonplace knowledge by now, yet strangely—and dangerously—many people still believe that when the body loses fluid it is important to restore salts to it. Workers in steaming hot factories are still given salt tablets by management-employed nurses as a way of combating the fatigue caused by work in those industrial infernos. Many people after vomiting or other kinds of illness will take something salty, thinking they are doing their bodies a favor. Unfortunately, what the body needs is something quite different.

After exercise, intense heat, or other conditions that deplete the body fluids, the body needs first of all to have fresh fluids and secondly to have its *potassium* supply boosted. Only in the rarest, most severe kinds of purgings does the supply of bodily salts fall below the necessary levels but, as we shall later discuss in some detail, our bodies waste valuable potassium while greedily hoarding sodium. Doctors believe that the body's tendency to store its sodium and its concurrent inability to retain potassium is a kind of "evolutionary lag."

Physicians are finding that more and more diseases, ranging from the annoying to the deadly, are attributable to the retention of sodium and fluids. An abnormality in the all-important fluid–salt balance can definitely stimulate a variety of disorders. Some of these disorders can be called psychological in that they affect our moods, our spirit, the way we see the world and the people around us. These so-called psychological problems are often the most painful and destructive because we who suffer from them tend to blame something in ourselves, in our characters, for the misery we undergo, and it often takes years of unhappiness before something is undertaken to change the situation. Other disorders are distinctly physical, and rank with some of the major killers of our time: hypertension and kidney disease.

Hypertension. Depression. A life of stress and frayed nerves. Do these sound like the complaints of some high-powered executive sitting behind a mile-long desk with fifty hold buttons on her telephone? Well, they could be, but in my case they were not. It was the story of my life and I was far indeed from some high-powered type. As a single woman in New York City I suppose I had my share of high-tension situations, and I went for a long while rationalizing my low spirits and chronic misery with such statements as "I worry too much," or "I guess that's just how life is." It would have seemed incredible to me a few years ago if someone had suggested that I could make truly remarkable, truly dramatic changes in my life just by severely regulating my salt intake. I simply had no idea that those little white crystals— which I had seen since infancy and which were everywhere from my grandmother's kitchen to the beautifully set tables in New York's poshest restaurants—were dragging my body and spirit into a perpetual rut of misery and dissatisfaction.

I myself was a chronic salt victim and didn't even know it.

CHAPTER THREE

Salt & Anxiety
A Personal Story

When I began thinking of writing a book that might
help warn others about the great dangers of excess salt,
it was my deep hope that even those who, as far as they
knew, weren't suffering from any of the terrible side-
effects of sodium overload would take heed. Too often
we wait to be truly sick before we begin to concentrate
on taking proper care of our bodies and then, quite often, it is too late.

KILLER SALT, then, is not a book just for those who already knowingly suffer from excess salt ingestion and it
is not even just for those who knowingly suffer from
one or two of the symptoms of salt overload. I have
meant this book for anyone who salts his food or eats
food with sodium compounds added to them. As we go
on in our discussion I will be able to outline in some detail how our bodies react to salt and just how salt plays
its often deadly role within us. And it is my hope that
by giving a clear picture of salt and the body I will be
able to help not only those who already suffer from salt
consumption but those who *may*, if their habits do not
change, suffer in the future.

Until a couple of years ago, I must admit, I was not
only a great abuser of salt but also was one who would
have—had anyone tried to inform me—turned a deaf ear
to the perils of sodium overload. I wish I could say that
my unwillingness to consider the dangers of salt came
from my feeling as fit as a fiddle. I wish I could say that
I ignored my health because my health was perfect, or

even good. But unfortunately this was not the case. Now, looking back, it is a little hard to believe, and more than a little embarrassing, how I allowed myself to feel miserable month after month; how I resigned myself to whole chunks of dead time and never sought to truly improve my situation. I had to wait for my own personal situation to get worse and worse before I began to deal with it.

I found it difficult to believe that such a common, and let's face it, such a tasty compound as table salt could ever possibly do me any real harm. It positively makes me cringe to remember how I avoided doing what was best for me and how, once I was on the road to living salt-free, I backslid—sometimes out of a strange desire to pretend that the obvious harm salt had done me was somehow "not real" and that I could return to my old ways.

The focal point of my salt-suffering was usually the week preceeding my menstrual period. If this seems like a "trivial complaint" to you that only means you have never experienced this kind of monthly agony. Perhaps my symptoms were more severe than the average young woman's but the truly astonishing fact remains that a full 40 percent of women in their "child-bearing years" suffer from premenstrual tension, and some will say that the percentage is even higher. I know that in my experience I can't remember even *one* female friend who has not had either regular or sporadic premenstrual discomfort, be it pains, depression, shortened temper, or disorienting weight gain.

From the very beginning of my sexual maturity I had associated my period with a time of discomfort. I happened to have been one of those girls who actually look forward to her puberty and I was lucky enough to come from the kind of home in which sexual and biological matters were discussed with relative freedom and openness. I welcomed my first period and looked on it as a kind of badge, a way of being grown up, a step toward real womanhood. That the week before each period was often almost crazily depressing struck me, at the time, as my body's way of getting used to its new capabilities and I just assumed that as I grew older and more used to menstruating the monthly discomfort would fade away.

It was in this way that I learned to *accept* feeling ill. Throughout college, during my first years on my own, through my first job, my first love affairs, travel and the attainment of many of my most ardent goals, I lived a life of three-week months, with one week regularly marred by premenstrual symptoms that ranged from feeling "down" to feeling painfully anxious. The fact that nearly every woman friend I had shared with me her own version of my story led me to believe that this lost week out of every month was just something I had to live with—if I could call it truly living. Once, in a particularly bad mood I took out a pencil and paper and did some depressing calculation. If I lived to be sixty years old and menstruated for thirty of those years I would have, figuring at the rate of one week lost out of every month, lost more than six years of my life. It all seemed so unfair.

I looked everywhere for reasons for my feeling poorly, for even though I accepted my terrible discomfort as my "lot," I couldn't help wondering if there were variables in my life that made it worse than need be. I told myself that once I landed the job I wanted so badly the tension would no longer come over me every month. So I landed the job and the tension and anxiety did not lessen at all. I told myself that if I could find a man I really loved and respected and could live with that man in a relationship of true sharing, then perhaps the tension would disappear, along with the other premenstrual symptoms. I found myself in a relationship with someone who really *did* live up to my dreams of happiness with a man, but unfortunately for me (and for Julian too) I could only really be in that relationship three weeks out of the month. For despite my happiness and satisfaction I was still plagued by monstrous depressions, unattractive sudden weight gains and fits of temper so severe that sometimes I felt I must be mad—for I truly didn't recognize myself in the midst of my anger and often, when it had passed, I looked back at my outbursts with a kind of sad astonishment.

It's not so terribly difficult for me to speak of it now but before I went salt-free I was growing ashamed of myself. Despite what I thought was a basically healthy and positive attitude toward my own body I was gradually beginning to think of my body as my enemy, as something that was almost alien to me and would attack me and the life I wished to lead.

Reaching The Breaking Point

Julian, the man of my life, is a successful lawyer. He also happens to be extremely sensitive to other people, particularly me. He was, of course, aware of the difficulties I experienced monthly and while he was certainly concerned and probably even disturbed by them he knew how to stick by me and always be supportive.

One month, just before I began my premenstrual decline, Julian's professional obligations took him to Japan for a week. I was nervous and jumpy to begin with and Julian's absence only worsened the situation. The first night of Julian's absence I tried to act as normally as I could but I found I couldn't even make myself a normal dinner. The sight of that little half chicken sitting in a roasting pan struck me as grotesquely sad and the sight of our table set for one brought a sudden gush of tears. I felt stretched to the breaking point. Quickly, I threw the chicken into the freezer and began a regime of quickie foods that was to last until Julian returned. That first night I ate a cube steak broiled in salt and butter and from there on I degenerated to peanut-butter-and-jelly sandwiches!

What I had traditionally noticed most acutely during my premenstrual horrors was how my mood was affected, but without Julian to react to (and against) I was more struck by the second part of my monthly misery—the puffiness that turned to absolute bloat. Now I am not one of those women who preen and want nothing more than to look like some emaciated high fashion model, but like most young women—like most *people*—I wanted to look attractive, both for others and for myself. In a couple of days I felt uncomfortably bloated and I used Julian's impending return as a kind of goal—I would lose the weight before he got back.

What I didn't understand is that water weight is caused by the body's *retention* of fluids. Unarmed with any real information, I tried to solve my problem by going on a low calorie diet consisting mainly of Chinese food, soy sauce, and as much water as I could pour down my throat. I had read in a couple of magazine articles and popular diet books that drinking lots of water is a good way to lose weight—something about flushing the excess pounds out of the body. During the day I was extremely busy so my lunch generally consisted of a few slices of meat—usually roast beef—dipped in salt. I thought I was being terribly wise. What could be more slimming?, I thought. Just protein, no bread, and salt to make it interesting. I simply had no idea that salt was the worst thing I could eat during my premenstrual cycle and no one had ever taught me that both Chinese food and soy sauce had loads of sodium in them.

And so my crash Welcome Back Julian diet was a total fiasco. Rather than losing weight, or even staying at the same level, I continued to gain! I felt literally trapped inside a body that was changing for the worse without my being able to do anything about it. The unwelcome physical changes exacerbated my poor psychological state and my poor psychological state doubtless worsened me physically. By the night before Julian's homecoming I was no longer looking forward to seeing him, or, that is to say, I no longer wanted to be seen *by* him.

Julian arrived home at seven in the morning and he was enormously glad to see me. Missing me as much as he had, had perhaps blinded him to my weight gain and to the terrible strain that showed in every aspect of my being—my face, my voice, the way I responded to his touch. He felt sad to be in such a rush—he had to be in court at 8:30—but I was glad he'd be going and that our reunion would be cut short.

"How was Japan?" I asked him as if he were a distant cousin.

"A lot of work," he said. "Look, I know I have to get to court and you've got to work today too, but couldn't we just relax together for a few minutes?"

"I *am* relaxed," I said, with an edge to my voice.

He smiled shyly. I was making him so uncomfortable. "I mean relax together. We could lie down."

I was wearing a kimono to cover the pounds I'd put on, for although I wasn't exactly fat, I felt ridiculously flabby. I ignored Julian's amorous advances, choosing to pretend I didn't understand what he meant.

"I'll make some coffee."

"O.K.," said Julian. As I said, he was used to my moods. "I'd love some coffee. Couldn't get a decent cup of coffee in all of Tokyo. I wonder what the Japanese businessmen find to complain about when they come to New York."

Oh, aren't you the chipper one, I thought as I trod off to the kitchen to make the coffee.

"I'll make muffins," said Julian, following after. "Are there any?"

"Yes," I said. They were bought three days before Julian's departure and they were still unopened.

And so we had our little reunion breakfast. We drank a full pot of coffee—one of those twelve-cuppers—and, starved for some good old starch, I had three English muffins, each half spread with a thick coating of bright yellow salted butter. I laid it on as if it were cheese. If we said anything to each other I don't remember a word of it.

Finally Julian looked at his watch and said he'd better change his suit and get downtown to court. I then remembered that I was out of cash and needed to borrow money for carfare to work. It was this innocent oversight that triggered the real fight. I asked him for a dollar.

He said, "Sure thing," and opened his wallet. In a plastic window which is visible as soon as the wallet is opened, Julian carries a picture of me. I was used to seeing it there and usually didn't pay any attention to it. This morning, however, I noticed it was all wrinkled and stained—a yellowish-orange stain.

"What's that?" I said, pointing to my disfigured likeness.

"Oh God, that," said Julian. "I'd already blocked it. One of the stewardesses spilled orange juice on it by mistake." He took the picture out of his wallet, looked at it with a kind of abstract pity, and then placed it under a giant piece of quartz we keep on our kitchen table for decoration.

As I watched him place a picture of me beneath a rock I felt my entire body heat up with rage. Not wishing to make a scene I quickly got up and went to my dresser to put on some makeup, leaving the dollar untouched on the kitchen table. Julian, with a kind of foolish bravery, followed me in and was about to ask what was wrong.

But I wasn't going to listen to *him*.

"So," I said, "what kind of little game were you playing with the stewardess? Why were you showing her my picture?"

"Oh no," he said, "you don't understand. It was hours ago. Somewhere over California. No, even before then. Over the Pacific. I bought a screwdriver from her and while I was paying, the plane took one of those little hops and she spilled the drink. Do you want to smell my wallet? The whole thing stinks of booze."

He sounded so sweet and so believable and I could feel, almost as a physical sensation, a part of me going out to him. But along with the good feeling came an increase of rage and it was the rage that won—rage and fear. I actually believed as I looked at him that he was lying and that he had been flirting with a stewardess over the Pacific. Yet at the same time I *knew* I was wrong. I became so confused I did something I very rarely do: I burst into tears. Through my sobs I heard Julian calling out to me but I literally ran out of our bedroom and locked myself in the bathroom. When I finally emerged, Julian was gone.

Later that day, after a perfectly dreadful day at work, I came home. Of course I didn't expect Julian to disappear because of a fight but when I found him waiting there I was wildly grateful. He was groggy from a day's work plus the jet-lag of a twenty-hour flight, but he'd forced himself to stay up to tell me, "You've got to help yourself. I try not to take these fights too personally but it's hard. And it must be a hundred times harder for you. It's just not fair that you should go through this."

"I've talked it over with my gynecologist," I said.

"This is a big city," said Julian, "and there are plenty of doctors. Talk to another one. What have we got to lose?"

CHAPTER FOUR

How Salt Causes Premenstrual Tension

I was lucky. I found a very sympathetic gynecologist. One who would listen. I explained in detail the feelings I went through before my menstrual periods and he listened carefully, without rushing me through my story. It may have been a very familiar tale to *him* but it was important to me to explain it in detail. Then he gave me a routine but thorough examination. I think I almost *wanted* him to find something wrong. Nothing terribly serious, of course, but something definite which could be easily corrected. But his examination revealed no physical abnormalities.

That's when he asked me the question that turned out to be the turning point in my life.

"Do you like salty food?" he asked.

"Well," I said, "I've never really thought about it. I suppose I do. I use it in cooking and of course I like to put it on my food. Why? Can I cure my condition with salt?"

"Hardly," my doctor replied. "I'd hazard a guess that salt is just the thing that is doing you in. My guess is you're having far too much salt."

"Too *much* salt?"

"You see, the one thing out of the ordinary I happened to notice during your examination is you're holding quite a lot of fluid in the tissue around your hips and ankles. Even your hands and fingers. Are you aware of that?"

I nodded. I was horribly aware of it and found myself blushing.

"If you pinch yourself there," he went on, "you'll feel that it's rather soft. That's the kind of thing that you

21

get with fluid retention. Now when you were describing your premenstrual symptoms you said you generally gain a few pounds during the ten days before your period. Is that right?"

"Yes."

"Well, salt causes us to retain water and that water is pressing on all the tissues of your body. Even your brain. And from what you've told me it's making you, well, a little short-tempered."

"To say the least," I said. We shared a laugh.

"O.K.," he said, "then it's making you a crazy person. Now look, salt will cause anyone to retain water. That's what it's for. Its function in the body is to maintain bodily fluids at the optimum level so that the blood is at the right consistency and the tissues have as much fluid as they require. In many people there is an inborn tendency to retain water. It *may* stem from a sort of supersensitivity to salt and it's not at all unlikely that someone else in your family shares this tendency with you.

"Now what I'm going to suggest is that you cut out all additional salt in your diet during the ten days before your period. You'll get as much sodium as you need just from eating normally—from things like bread, meat, and cheese."

"But I *have* to eat salt on some things," I said. I hadn't really thought about it before but now that the doctor was suggesting I stop salting my food I realized I had a real craving for the stuff.

"That's just what you've been taught," he said, "and it's what you've gotten used to. But there's no reason why you have to use any salt on your food whatsoever. I know a lot of people on salt-free diets and they'll all tell you that in a short while you won't miss the taste at all."

Q. *Why Don't All Women Know How Salt Affects Their Menstruation?*

There have long been prejudices against menstruating women. Our primitive ancestors, unable to understand the function of menstruation, viewed the blood with fear and revulsion. Some argue—convincingly at times— that the various taboos against the menstruating woman were a part of the psychological warfare in man's quest to dominate and subjugate woman. Be that as it may, the fact remains that many of the ancient prejudices against menstruating women have survived and are with us today. Women are still thought of, by women themselves as well as by men, as somehow "unfit" for positions of high responsibility because of the much ballyhooed "raging hormonal imbalances" that supposedly accompany menstruation and which supposedly cause women to be unable to control themselves. The simple fact, of course, is that the reproductive cycle in women is not a matter of hormonal "imbalances." Rather there is a cycle of perfectly elegant balances and shifts which are still not completely understood by the medical profession. (One of the great contributions of the Women's Movement in the Seventies is a growing bibliography dealing with the quantitative and qualitative facts of menstruation.)

For a long while the millions of women who suffered, as I suffered, from premenstrual problems were "poohpoohed" by the medical profession and told to have more "self-control." In the light of what we are now learning, that premenstrual problems are a physiological dysfunction, this kind of cavalier treatment of suffering patients is disgraceful. While the proper subject of this book is not the various and harmful misconceptions we have held about menstruation, it is important, I think, to touch upon this subject. Among other things, it shows, I think, how so many can hunt so far afield for explanations of phenomena that can often be explained and remedied dietarily.

In reading Paula Weideger's excellent and valuable book *Menstruation And Menopause*, I came across the opinion of one of America's most renowned psychiatrists, Karl Menninger. According to Menninger, premenstrual depression results from woman mourning her failure to conceive that month which then leads to a rejection of her feminine identity.

> "The envy of the male cannot be repressed and serves to direct her hostility in two directions: she resents the more favored and envied males while secretly trying to emulate them, and at the same time she hates and would deny her own femaleness."

I am not antagonistic to psychoanalytical theories. Freud was unquestionably a man of courage and deep brilliance and many of his followers have also made tremendous contributions to the understanding of human life. But there remains a terrible suspicion when one reviews the work on menstruation done by these often brilliant physicians. Is it far-fetched to believe that a woman could complain to a Dr. Menninger, or even a Dr. Freud, of her terrible premenstrual depression, and even as she spoke be eating a bag of salted nuts—and still be given no simple, biochemical explanation for her symptoms? It is precisely *here* that the age-old myths and terrors surrounding menstruation have delayed our true understanding of the body.

I am not trying to suggest that all of the studies relating to menstruation and premenstruation be scrapped and that all of the discomfort be attributed to excess salt. But isn't it time that women stop being told—and stop believing—that the various changes they experience during their menstrual cycle are a function of their psychological makeup alone and cannot be dealt with easily and successfully physiologically?

Q. *How Does Salt Affect The Premenstrual Cycle?*

Estrogen and progesterone, two of the principal female hormones, are in an intricate balance throughout the month. When one is at its peak in the body, the other is then at its lowest level. (This crucial cycle is regulated by our pituitary gland.) After ovulation, after the ripened egg is released, the levels of estrogen begin to taper off, and progesterone begins to increase. Now, recent research suggests that of these two hormones progesterone has the most influence on our sodium-fluid balance. *Any* salt taken into the body at this time will be, therefore, more easily reabsorbed by the kidneys and due to the influence of the progesterone we will then find ourselves retaining higher amounts of fluid.

And what is the result of this? There seems to be little doubt. The effect of the excess fluid pressing upon the tissues of the body (including the brain) is one of depression and a feeling of tenseness. While we have long thought of such things as feeling low, feeling snappish, or feeling just plain mean as purely *emotional* phenomena, it now seems certain that these "moods" can be brought on by excess fluids in the body—which, as we now know, can be caused by salt.

I know that in my case, the drastic reduction of salt around the time of menstruation (the beginnings of my salt-free life) meant remarkable and, I must admit, astonishing changes in my life—not only in those once "bad days" but in the whole month, for I no longer dread the oncoming menstrual period. And as I began to study ever more closely the effects of salt in general, I became convinced that thousands and thousands of women could unquestionably be helped by greatly reducing their salt intake.

My research has convinced me that cutting down on salt is not likely to make an enormous difference in every single case of premenstrual problems. In the next chapter we will have a good look at salt and how it relates to the current medical concepts of *stress*, but for the time being let us note that some observers attribute some premenstrual discomfort to stress, which means it is due to a confluence of factors, operating in a kind of vicious circle, each feeding back to the other. With this in mind, observers, some of whom readily accept the role of salt and fluid retention in determining mood, point out that fluid retention can be affected by psychological and cultural factors.

Q. *What About Sugar?*

The highly volatile emotions preceeding menstruation have also been attributed to low blood sugar—hypoglycemia. During the premenstrual cycle, some say, spontaneous hypoglycemia can occur which can cause nervousness and distorted thinking as well as nausea and an abnormal craving for sweets. Despite what some people say about the body innately knowing what is best for it, this craving for sweets is not a good way to handle the problem of spontaneous hypoglycemia. The best thing to do is stick to proteins. Loading up on sugary foods is never a good idea. They are burned up right after ingestion and though they do cause a sudden, initial rise in blood sugar they let you down just as quickly. In order to best defend yourself against fluid retention and hypoglycemia, physicians advocate a low salt, low carbohydrate diet.

Q. *What About Salt And The New Studies of Mood Cycles?*
A very recent theory about premenstrual syndrome is
that it is a cyclical disorder, like manic depression. Manic depressives are characterized by their abrupt and extreme swings of mood, going from the depths of depression to wild, often frightening euphoric heights. This
disorder is cyclical and once the length of the cycle is
determined it is possible to some extent to predict when
those who suffer from it will hit their emotional highs
and lows. Leaving aside the question whether premenstrual syndrome can (or even should) be compared to a
serious ailment like manic depression, it is intriguing to
learn that some of the most astounding breakthroughs
in the treatment of manic depression have been based
on medical science's new understanding of salt in the
body. This new understanding of the manic-depressive
cycle is connected with the development of what some
believe to be a wonder drug in its treatment—lithium.
Lithium is a naturally occurring element and was at one
time used as a salt substitute. Although researchers are
not yet entirely certain how lithium works in the body,
it is believed that it affects the body's mineral and fluid
balance. One of the pioneers of lithium use is Dr. Ronald
Fieve. In his book, *Moodswing, The Third Revolution in
Psychiatry*, Dr. Fieve writes:

> In manic depression mood shifts are accompanied
> by shifts in body chemistry, particularly in the
> amounts of salt and fluid in and around the cells.
> Depressed patients studies . . . have consistently
> shown that they retain salt and fluid only during
> their depressed phase. Lithium's primary action is
> on salt.

Although lithium is rarely used in the treatment of
premenstrual syndrome, it does help all of us to look at

the disorder as a cyclical disruption of our fluid balance. With this in mind, there is, I believe, a lot you can do to help yourself, without blaming your psychological makeup, without resorting to drugs. If premenstrual sufferers would only cut down on salt (and also drink a little less fluid) during the time of premenstrual stress it is virtually certain that a difference would be felt. Not only in the long run but *right away*.

It isn't difficult to cut down on salt, but, as we shall demonstrate later on, it is not always easy either. This isn't because salt is so necessary to make food taste good. If it were only a matter of the salt we ourselves added to our foods we'd have an easier time taking care of ourselves. (We'll also talk about how to make food taste really delicious without a grain of salt.) But unfortunately our large food manufacturing corporations see fit to add unspecified amounts of salt and other sodium compounds to our foods and so those who eat canned or processed foods may be getting severely overloaded with salt without ever touching their salt shakers.

The only guarantee is to eat fresh foods—foods naturally low in sodium. (There's a sodium content table at the back of the book). I urge you to try this low sodium cure. Surely, if your discomfort continues it would be wise to take further action. But no matter what sort of doctor you share your problem with, don't let *anyone* tell you that it's "all in your head."

We know too much to listen to that kind of thing anymore.

CHAPTER FIVE

Salt & Stress

The word *stress* is a common one in our everyday language, but stress is also one of the most important medical concepts of this century. Stress plays a crucial role in nearly all of our modern illnesses. We know now that salt, when its ingestion is combined with stress, will lead to one of the terrible, killing diseases of our time— hypertension. When I first learned how salt was turning my premenstrual cycle into a premenstrual nightmare, I thought I had learned the worst effect of our salty diets, but as I learned more about the concept of stress I realized that the violence committed by salt against our bodies was, in fact, far graver than I had feared.

The changes produced by stress are intimately connected to the body's use of salt. The body's reaction to stress can stimulate endocrine hormones which will cause both salt and water to be retained. We have already seen how the process causes depressing and often overwhelming mood swings. But the chilling fact is that it also ravages the kidneys and the entire cardiovascular system.

Q. *What Is Stress?*

The discovery of the stress syndrome, properly called the General Adaptation Syndrome (GAS), is credited to Dr. Hans Selye, who has won the Nobel Prize for his breakthrough discovery of the GAS. His landmark book, *The Stress of Life*, explained the evolution of the stress concept and how it relates to disease. Dr. Selye's General Adaptation Syndrome was an early attempt to unify the disease process into one theory which would be universally applicable.

Selye was the first to explain that stress could be caused by such diverse (and overlooked) factors as infection, noise, and overcrowding, or tumultuous but usually pleasing events such as marriage or a sudden rise in one's financial status. Adaptation is the intrinsic factor in all of these seemingly diverse stressors. As we well understand now, *stress* turns out to be a factor in all disease. And, interestingly, the reactive physical changes engendered by stress are the same no matter what the source of the stress or what organ is involved.

We now know that many of the diseases which so plague us today are explainable, at least in part, by the stress syndrome and we are also learning more and more about how salt intake can affect this potentially deadly syndrome. I think it's vitally important that we better understand stress since it appears to be a factor in practically all disease.

The Three Stages Of Stress

According to Dr. Selye, there are three stages in the stress syndrome and each of them is distinct, no matter what is causing the stress. The stages are: alarm, resistance, and exhaustion. Each of these three stages is associated with specific bodily changes and each stage involves the same organs each time. It is one of the remarkable characteristics of stress that the body reacts the same way no matter what *causes* the stress.

Stress, in this framework, means anything which necessitates an adjustment. For instance, if you happen to suffer a minor household burn, a localized mechanism called the Local Adaptation Syndrome is activated (LAS). This LAS is linked to the process of, in the case of the burn, local repair of injured tissues. The GAS and the LAS are closely linked and the identical physical defenses are set up in the body by both of them. Both

GAS and LAS involve the release of a pituitary hormone which stimulates the adrenal gland to secrete the hormones necessary for the defense and/or repair of the body as a whole.

The physical changes that are a result of *stress* are the body's attempt to somehow adapt to the cause of stress, be it noise, grief, or extreme cold. In this respect, our bodies pay almost as high a toll for so-called emotional stress as they do for physical stress, because the changes brought on by either are identical. And what is the ultimate risk of extreme and continued stress? Nothing less than death. At a certain point in its fight against stress—such as that minor household burn—it may mean the death of just the affected tissues but in larger, more pervasive stress the final result may well be the death of the entire organism.

Stress And The "Working Girl"

Because the theory of stress is so crucial to our understanding of salt—and our understanding of how we may literally be poisoning ourselves to death through our uninformed use of salt—I think it will be helpful if I share an actual stress situation (from my own life) and explain how it affected me internally. Though the three stages of stress are distinctly physiological it will be clear that the psychological factors are an important part of how hard the stress hits the body. Dr. Jay M. Weiss of the Rockefeller University, a leader in stress disease research, feels that the more helpless you feel in a situation, the more likely you are to develop stress diseases. As more and more doctors are learning, the person who learns the secret of remaining calm, but who doesn't swallow or repress feelings, will stand a far greater chance of avoiding the many ailments caused or worsened by stress. Unfortunately, however, even those physicians who now warn their patients about stress and its dangers very often do precious little to reeducate the patient about basic dietary truths. In other words, it seems a little self-defeating to warn someone about heart disease or high blood pressure and not also warn them about the role excess salt has in worsening these stress-related diseases.

As I describe in some detail the impact of stress in my own body I will be using a certain amount of medical language. I don't think I will be using terminology that is too technical, but I do want to remind readers of the glossary at the back of the book.

Like so many before me, before I found a job that I could truly be happy in I put in a lot of time at a lot of nothing jobs. I, of course, needed to make money and I told myself—when I needed cheering up, which was often—that I was also gaining a lot of valuable experience. Well, I might have been gathering experience but I was also (following a family tradition) building up a lot of terrible tension. I was working in a crummy job for a boss who was himself a high risk type for stress diseases—irritable, harried, and as tense as a drumhead. I hated facing him and my dull job and no day was worse than Monday.

A typical Monday. Up at the alarm clock's wretched shriek. Quick (and salty) breakfast, which I can chew already thinking about where I have to be going—and the rage is on its way. I pop a few salted crackers in my mouth to "calm my stomach" and then submit my already tense system to the very real assault of the subway. The one-hundred-decibel screech of the underground train is, in itself, enough to alarm my system, but the whole stressful situation has been made worse, first by my salty diet and then by the man who sits next to me on the train who manages to nudge me with his elbow each time he turns the page of his morning *News*.

My jaw is tight by the time I arrive at work and then, naturally, I find that the elevator is, as the superintendent likes to put it, "on the blink." I feel like decking the old gentleman, but of course a lady doesn't behave that way! So, swallowing my annoyance, I begin the climb up the six flights of unswept stairs. This last annoyance could have done me some good—after all, we urbanites need our exercise—but the rage and general psychological stress engendered by the forced climb pretty much cancel out its benefits.

My boss is waiting for me. He's been trying to, as he puts it, "trim down," which really means that if he doesn't lose forty pounds his doctor is going to refuse to see him any longer. To facilitate this weight loss, my boss quiets his appetite by sipping bouillon practically all day long. It hasn't worked at all. And he doesn't know—and at that time neither did I—that each bouillon cube is packed with one of the more debilitating sodium compounds, monosodium glutamate (MSG). Holding his little plastic cup of beef bouillon, he eyes me for a moment while I catch my breath.

"Have they fixed the elevator yet?" he asks.

"No," I manage to say through my panting. I wonder how he can ask such a stupid question.

"Well, that's just too bad," he says, suddenly annoyed. "Don't bother to take off your coat. I need you to run over to the purchasing department and pick up ten pounds of powdered lactose and then deliver it to Dr. Rhodes in the pediatrics research unit. Then I want you to come back here and deliver those back orders from yesterday. And you might have to go over to the supplier in Jersey today if he's got time to see you."

By the time my boss was finished with me I was well into the first stage of the stress syndrome, the stage of alarm. My body had been alerted to impending harm—whether real or imagined—and it was mobilizing itself for defense. Outwardly I may have appeared calm but internally things were really moving! My hypothalamus was sending frantic hormonal messages to my pituitary gland in response to the stress.

Q. *How Does The Brain React To Stress?*

The hypothalamus is a deep-seated part of the brain. A section of the hypothalamus is responsible for the control of two all-important human functions: instinct and emotion. When people talk about "gut-reactions" they are more often than not talking about responses which originate in the hypothalamus. The hypothalamus also controls some of the actions of the gut such as eating and drinking. The hypothalamus is in the most "primitive" part of the human brain, the deeply buried part which is as old as the race and not subject to the refinements of rational thought which belong to the cerebral cortex, a comparatively recent addition to the brain.

The messages from the hypothalamus reach the body via the front lobe of the pituitary gland. This small and vital gland finds its home on the floor of the brain cavity near the hypothalamus and, like the hypothalamus, it is part of what can be called the brainstem, parts of the brain which were evolved early in human history. When the pituitary is stimulated by the hypothalamus it means energy is needed for such physical emergencies as fighting or internal repair. The pituitary, as soon as it receives the hypothalamus's message, sends out a hormone called ACTH. This ACTH goes to a part of the adrenal gland known as the adrenal cortex. ACTH acts exclusively on the adrenal cortex, which finds its home wrapped around the adrenal gland above each kidney.

With these messages flying around, your body is preparing for what has been called "the fight-or-flight response." The blood sugar increases, the pupils dilate (look in the eyes of a frightened person—the eyes themselves may enlarge but the pupils dilate), the muscles contract, and the blood pressure rises temporarily. As if hedging its bets, while the body prepares to make its stand, it also releases certain substances which aid the cells in repairing themselves and these substances are immediately discernible in the blood stream.

Of all the substances the adrenal gland releases, one of them is most closely linked with the feelings of anger and tension. This substance is called norepinephrine (NE) and it is secreted by the inner portion of the adrenal, the medulla. NE stimulates the central nervous system and it is this trigger action that we experience when we are feeling angry. In fact, an injection of NE into the brain will cause anyone to feel anger, no matter how peaceful they were the moment preceding the injection. If you are unfortunate enough to be secreting this substance in excess you will be feeling tense and angry an enormous amount of the time. NE acts to constrict—to tighten—the blood vessels and it also acts in conjunction with another hormone from the adrenal cortex, aldosterone, to raise your blood pressure. It is here that we can see the potentially deadly effects of salt consumption and how excess salt can turn anxiety into a very real and deadly physical malady. The aggressive energy liberated by the chemical changes of the fight-or-flight response is usually given no outlet in modern man's life. Whereas formerly the energy would have been used for battle or flight from wild animals, now there is no outlet for these emotions behind a desk, and the chemicals are left boiling around, keeping the body in a state of constant preparation for battles which usually are never fought. This overstimulation of the adrenals is a by-product of our modern, stress-filled lives. It keeps the body in a state of continual tension without release, and it also changes the body's response to excess sodium, making it more easily retained.

Aldosterone acts to retain salt and fluids. So as aldosterone, in response to ACTH and hormones from the kidney, causes us to store up salt and fluid, NE, as we have already seen, causes our blood vessels to constrict. The result is, quite logically, high blood pressure. Given aldosterone's natural tendency to preserve fluids in time of stress, we are wildly increasing our risks if we also dose ourselves with salt. For salt, as the meat packing industry has always known, is a sure-fire way to retain water.

Q. *How Do Salt And Stress Affect The Kidneys?*

One of the major parts of our body to pay the price for this salt- and hormone-triggered reaction are our generally vulnerable kidneys, which reabsorb the sodium and then pump it back into the blood—a real vicious cycle since once the sodium is pumped back out it is again reabsorbed by the kidneys.

My study of salt, by now, had taken me some distance from my original complaints. Now I was face to face with one of the terrible diseases of our time—high blood pressure.

Q. *Why Does Our Blood Pressure Rise?*

A rise in blood pressure is usually caused in one of two ways. Either the walls of the blood vessels grow smaller or the amount of blood in circulation increases. One of the ways our blood supply increases is through the introduction of increased water into the blood stream. So you see one needn't travel very far in the study of high blood pressure to be very directly confronted with the dangers of excess salt. This increase in our blood pressure, when it is triggered by stress, lasts as long as the stress lasts. (In my case, afternoons were generally easier than mornings, because I could feel the end of the work day approaching and I relaxed.)

Had I been undergoing a more major stress eventually I could have reached the stage of exhaustion which can come about if the centers pumping out the adrenal solutions become exhausted. If this should happen you would be facing a serious medical problem. When the adrenals become exhausted—as they can—and stop producing aldosterone, a great deal of the body's natural and constant supply of sodium is suddenly lost and the blood pressure, rather than being too high, drops off and can then be *too low*. If this should happen the body can literally *die* from low blood pressure or heart failure. (In another chapter we'll discuss why the body must have sodium.)

Q. *Are The Effects Of Salt and Stress Just Temporary?*

Every body is a map of the adaptations it has made to all of the stress it has encountered during a lifetime. My bad day at work, combined with a salted diet, was not quite enough to do me in, but as my system that day went through the stages of alarm and resistance, and as I heightened the entire destructive process with a diet that promoted fluid retention, I was leaving everlasting traces of that day in my body, just as emotional traumas leave everlasting traces in our character.

Medical researchers have taught us that stress is apparent on the internal organs, particularly on the adrenals. All of us, even hermits and children, carry the scars of adaptation to stress. Inside the most vivacious society hostess or the most spaced-out mystic are the traces of the body's reactions to diet, injury, and the world in general. According to Dr. Selye, nobody dies of what is called "old age." Rather, we lose our ability to adapt to stress any longer. And instead of bending—we break.

Thomas H. Holmes and Richard H. Rahe, two psychiatrists at the University of Washington Medical School have made a scale called the Social Readjustment Scale which rates the stressfulness of life situations. According to their reports as well as Selye's work, any drastic upheaval whether for better or worse is stressful in that it involves adaptation.

FROM HOLMES AND RAHE:
THE STRESS OF ADJUSTING TO CHANGE

EVENTS	SCALE OF IMPACT
Death of spouse	100
Divorce	73
Marital separation	65
Jail term	63
Death of close family member	63
Personal injury or illness	53
Marriage	50
Fired at work	47
Marital reconciliation	45
Retirement	45
Change in health of family member	44
Pregnancy	40
Sex difficulties	39
Gain of new family member	39
Business readjustment	39
Change in financial state	38
Death of close friend	37
Change to different line of work	36

EVENTS   SCALE OF IMPACT

Change in number of arguments with spouse 35
Mortgage over $10,000 31
Foreclosure of mortgage or loan 30
Change in responsibilities at work 29
Son or daughter leaving home 29
Trouble with in-laws 29
Outstanding personal achievement 28
Wife begins or stops work 26
Begin or end school 26
Change in living conditions 25
Revision of personal habits 24
Trouble with boss 23
Change in work hours or conditions 20
Change in residence 20
Change in schools 20
Change in recreation 19
Change in church activities 19
Change in social activities 18
Mortgage or loan less than $10,000 17
Change in sleeping habits 16
Change in number of family get-togethers 15
Change in eating habits 15
Vacation 13
Christmas 12
Minor violations of the law 11

Q. *Aren't The Effects Of Stress Inevitable?*

As far as we can tell, no dietary precaution and no
system of better living can shield us from the accumu-
lation of those emotional and physical experiences that
eventually wear us out. But it makes no sense whatsoever to rush toward our deaths full speed ahead. It
makes no sense to prolong anxiety-producing situations,
whether they be a style of life, an irritating job, or a

killing diet. Today, with so many of us leading lives of stress and subsequently suffering from its physical and psychological by-products, a certain level of discomfort is considered to be more or less normal—par for the course. It is, we are told, how we live now.

But it is important for us to know that it need not be that way. It is important to know that so much of modern life seems to lead to the deadly internal disorders we have been describing. The competitiveness and fearfulness of our society, our general lack of sound information about our bodies, and the incredible salt/sodium dosage that the American food industry continually pumps into our diets—all of these seem to drag us down into a state of discomfort and imperfect health. And sometimes, of course, this imperfect health is just the beginning of a tragic and premature decline that leads to death.

But just as I learned, with my doctor's help, that my years of premenstrual misery could be put behind me and that I need never suffer in that way again, so I learned—and we all can learn—that the way we suffer from the stresses in our life need not be so destructive. Isn't it incredible when you think about it that many of us believe that just because we happen to be alive in the twentieth century we have no right to feel truly and one hundred percent alive?

Q. *Is Salt Addictive?*

My old boss, who so bedeviled me back then, is probably still caught in his high salt diet and his lifestyle of total stress and tension, though I truly hope he has recovered. (A copy of KILLER SALT is going straight to him.) And there are, as we all know, millions like him. Living on the edge. Battling their way through each day, living lives of fear and crisis, and feasting on the salty foods that so terribly worsen the whole loop of the stress syndrome.

One of the true horrors of salt is that it becomes a kind of addiction—an addiction that strikes those who suffer most from its effects. It has been demonstrated that just those people who need it least crave salt the most. In a hospital study not long ago, doctors tested some hypertensive patients. Using some patients with normal blood pressure as a kind of control group the doctors gave the patients a choice of two kinds of drinking water. One was just normal fresh water and the other had a distinctly salty taste. Common sense would tell us that all of the subjects would have chosen to drink the fresh water but the fact is that the overwhelming majority of the hypertensive patients—those, in other words, who were already suffering from their ingestion of excess salt—chose to drink the salty water.

The significance of this experiment is as clear as it is frightening. For many suffering from hypertension, salt is the poison the body learns to crave. It is sobering to realize that in the case of salt the impulses of our body may be in direct opposition to what is good for it. In the most real sense, the body that asks for more and more salt may be, in a gradual yet certain way, plotting its own death.

CHAPTER SIX

Hypertension:
The Silent Hit-Man Of Disease

My study of salt eventually brought me face to face with my grandfather's sudden death. Early on, as I tracked the effects of salt, I suspected that this common substance had something to do with my grandfather's fatal illness, but the thought was, I suppose, too painful and I tried continually to put it aside. To think that he might have preserved his health and prolonged his life if he'd understood the effects of salt cause in me a kind of rage and frustration.

My grandfather died from the secondary complications of hypertension. He *knew* that salt was somehow "bad" for him, but he didn't really know *how* bad. His doctor had warned him to cut down on salt and my grandfather, in a manner unfortunately typical of him, pleasantly agreed, all the while paying the advice little attention. Even after suffering a severe stroke, my grandfather continued to place his doctor's advice outside of his daily life. As soon as he was able to my grandfather returned to work and, while he once in a while mentioned how he had to "watch it from now on," he never cut down on salt.

I remember one Sunday dinner. We were having a roast chicken dish my grandmother, a truly fine cook, had gone to great pains to prepare. She, too, thought my grandfather had to "watch it from now on," and she made a point of adding very little salt to her cooking. In truth, her meals needed no salt whatsoever. They tasted fresh, well-prepared, and she knew a great deal about

42

herb and spice combinations that gave her food a flavor far more delicate *and* enjoyable than salted food could ever achieve.

But my grandfather was, I now understand, a salt addict. Like many hypertensives, he had a deep craving for the salty taste. When his meal was put before him, he liberally dosed it with salt, before even tasting it.

"You know what Thomas Edison said?" I asked.

"Well it depends when you mean. What he said in the morning? What he said when he slipped on the ice?"

"No, I mean what he said about salt."

My grandfather still had the salt shaker in his hand and he was still dosing his food with the innocent white poison. "O.K.," he said, "let's hear it."

"Edison said that he'd never hire a man who salted his food before tasting it."

"Oh well," said my grandfather with a big sarcastic show of relief, "I'll remember that when I'm hiring new boys. Fortunately I no longer have the problem of seeking employment," he added, dusting his chicken with more salt.

Q. *Does Salt Cause Hypertension?*

Let there be no doubt about it: salt plays a major role in hypertension. Hypertension is one of the major maladies of our time. An estimated 25 million Americans suffer from hypertension. It is a killer in a great number of cases. Even though doctors now have sophisticated ways of treating high blood pressure more and more people are being afflicted with it; more and more are dying from it.

Salt and hypertension create a tragic, vicious cycle, with one contributing to the other. Clearly salt worsens and even causes high blood pressure, and now we have learned that those prone to high blood pressure seem to have a greater craving for salt. Tests have shown that hypertensives have a higher taste threshold for salt—that is, they need a good strong dose of the stuff to even taste it. The same study, conducted by a team of doctors and dentists, found that hypertensives also had lower sodium levels in their saliva than those with normal blood pressure.

We mentioned earlier a study conducted by Dr. Paul Schecter, David Horowitz, and Robert I. Henkin of the American Heart and Lung Institute. Hypertensives hospitalized for a week were fed a dry diet with a choice of distilled water or slightly salty water and four times as many of the hypertensives chose the *salty* water as the non-hypertensive control group.

Q. *Can Salt Help Us Diagnose Hypertension?*

As Dr. Henkin tells us, "Hypertensives have a definite and compelling drive to eat salt." He feels that this exaggerated appetite for salt could be a *symptom* of high blood pressure as well as an important contributing factor. He goes on to suggest that salt craving might be used as a simple screening test for unsuspected hypertension. As Dr. Henkin says, *"It would be a good idea for anyone who piles salt on his French fries to have his blood pressure checked."* This is clearly important advice. You owe it to yourself and everyone around you to honestly judge if you're a salt addict. If you are, or even if you think you *might* be, please see a doctor immediately. Often it isn't too late.

Most of us don't understand blood pressure. This confusion is understandable because the functioning of the heart, kidneys, and adrenal glands are very complex. Unfortunately, most of us aren't even *aware* of blood pressure until it goes up too high or happens to drop down too low.

In order for your blood to circulate through the *miles* of arteries and veins that circumlocute the body, your blood has to be under a certain pressure. This pressure functions as the locomotive force behind your blood movement. Your blood, of course, has to circulate in order to do any good. It carries oxygen and the nutrients required for life by your body's tissues. It also serves to conduct messages from the body's glands by carrying hormones. Doctors describe hormones as "glandular messengers."

The largest blood vessels are called arteries. Your arteries carry blood rich in oxygen away from the heart and lungs to every part of the body. Arteries divide into smaller units which are called arterioles and the even smaller capillaries. Arterioles and capillaries are too small to be seen by the naked eye. (The muscular walls of arterioles can raise blood pressure by contracting.)

The blood's vital nutrients pass into your tissues via the tiny capillaries. In the event of *stress*, more oxygen and such nutrients as sugar are quickly required by your body. In order to get this circulating as *quickly* as possible, your kidneys and your adrenal glands both release substances which cause the walls of the arterioles to contract. This contraction pushes the nutrients through the body much more quickly than when the body is functioning normally and without stress. And, of course, this increased pressure is measured as high blood pressure, or hypertension.

Q. *What Does Your Blood Pressure Mean?*

When you go to the clinic or a private doctor and your blood pressure is measured, your doctor uses a pressure-sensitive cuff known as a sphygmomanometer. The sphygmomanometer measures your blood pressure by registering how high a column of mercury rises in response to the pressure exerted by the beating of your heart. You are given a double figure, representing the systolic pressure over the diastolic pressure. The systolic pressure equals the pressure with which your blood is forced out of your heart; the diastolic pressure is the resting pressure between pulses. We used to believe that the diastolic pressure was more indicative of present or impending danger than the systolic, but now it's understood that a high systolic reading can also be deadly. A high diastolic reading tells us of one of the real physical calamities of hypertension—that the pressure within the whole cardiovascular system remains abnormally high even during the resting phases between pumps. In terms of the wear and tear on the system it is as if every day ages you two days. The normal range for blood pressure is usually around 120 (plus or minus 20) for the systolic and 80 (plus or minus 10) for the diastolic.

Relatively inexpensive sphygmomanometers are now for sale. In fact, on a recent country weekend I noticed one for sale in a large sundries store in Vermont. However, I haven't inspected these home devices and cannot recommend them. One thing is certain, however; you are toying with your life if you don't immediately seek a doctor's guidance and follow her or his instructions with great care.

Q. *What Causes Hypertension?*

It is now apparent that hypertension is due in part to some abnormality of the kidneys. This is so even if no chemical or structural change in the kidneys is visible. This new knowledge is extremely significant in furthering our understanding of this dread malfunction. Before, it was thought that the kidneys didn't become involved in a hypertensive ailment until it was malignant.

Malignant hypertension is, as the name implies, the most serious kind of hypertension and it means that the internal organs have been demonstrably affected—that is, deteriorated—by the disease. *Essential* hypertension means that the causative factors are as yet unknown and that the course of the disease is relatively a slow one.

In his study of essential hypertension, Dr. George Perera determined that patients with untreated essential hypertension had a lifespan of approximately twenty years from the onset of the disease. The uncomplicated phase of the disease, that is the phase in which organs such as the kidneys, the retina or the heart remained relatively sound, was as long as fifteen years. When the organ damage did occur, it was found that 74 percent of the patients studied suffered cardiac complications while 42 percent were stricken in their kidneys. More than half of the patients eventually died of heart disease. Ten to fifteen percent died of cerebral complications and ten percent died of kidney failure. The average age of onset of essential hypertension was thirty-two years. The mean age at death was, tragically, fifty-two years.

Q. *Is Hypertension Common?*

It is difficult to say with any certainty of accuracy just how many Americans suffer from hypertension. Some estimates run up to 25 million. *Low* estimates put it at about 10 million. Of those with essential hypertension, some will go on to develop malignant hypertension, but just how many is not known. What *is* known, however, is that untreated malignant hypertension can and often does lead to a fatality in roughly one year.

It is safe to say that at this very moment millions of Americans suffer from hypertension that has not yet been diagnosed. In this way, unlike other virulent diseases, hypertension is a hidden malady—hidden from the statisticians, and hidden from those who suffer from it.

It was not so long ago that high blood pressure was an ailment associated with hot-tempered tycoons and something that the so-called "common man" did not really have to contend with. It's true, of course, that business executives who live with a great deal of stress *are* targets for hypertension, but now we know that they are not the only targets and not even the *prime* targets. Hypertension, it is sad to say, shares a characteristic with most of the other killing diseases of our time: it is thoroughly democratic.

Hypertension undermines your entire body and makes true joyful, healthful living impossible. It weakens the heart, it probably affects its victims psychologically, and it is a number one cause of kidney failure. Physicians and medical researchers now agree that, though the exact relationship between hypertension and kidney disease is not fully understood, there is no question that the one can cause the other.

Q. *Does Salt Affect the Kidneys?*

Excess salt is such a major factor in high blood pressure, I learned, because of the kidneys' role in hypertension. Ideally, the kidneys ought to be able to handle just as much salt (or sodium) as we ingest, excreting it at a normal rate and never allowing it to accumulate in our bodies. However, this "ideal" kidney is not what millions and millions of Americans have working for them and in those cases the excessive sodium *cannot* be processed by the kidneys. Then there is a resultant fluid build-up which, naturally, increases the volume of the blood. This increased blood volume is what salt-induced high blood pressure is all about.

The kidneys' main function is to maintain the extra-cellular fluids and the blood level at an optimum level. If for some reason—say, kidney damage or genetic pre-disposition to salt retention—the kidneys don't excrete enough sodium, then serious problems will arise. Although hypertension research is still in its infancy, it seems the mechanism goes roughly like this: An excess of retained sodium increases the blood volume which in turn increases blood pressure. Also because of higher blood volume, the rate of heartbeat increases. Not only is the heart beating too hard and fast for its own good but the blood it is pumping through the body is of too great a quantity for the nourishment of the tissues. By a process of *autoregulation,* then, the blood vessels *con-tract* to reduce the blood flow. So you see the higher blood volume produced by salt causes, first, the heart to beat faster and then the onrush of blood causes the blood vessels to contract—which leaves you with hyper-tension. It is simple, automatic, and deadly.

The kidneys are your most crucial long-term regula-tors of blood pressure. If blood pressure is *low* the kid-neys respond by secreting a hormone called *renin.* Renin combines with a group of proteins in the blood to form a hormone group called the angiotensins. One of the hormones, angiotensin II, is now thought to raise your blood pressure by causing the walls of the arterioles to contract. Angiotensin II also stimulates the release of al-dosterone from the adrenal cortex. As we discussed in the chapter on stress, aldosterone acts on the kidney to increase the rate at which it can reabsorb sodium. Al-dosterone acts directly on sodium and fluid levels to raise the blood pressure by expanding the blood volume.

If you happen to be lucky enough to be innately im-mune to high blood pressure then the chances are that when your blood volume increases it will not remain high. However, if you have some malfunction of the kidney—either known or undiagnosed—or have a family history of hypertension, then the chances are that the high blood volume created by the kidney will be maintained and that critical hypertension will be the result. Medical researchers are coming to believe that each kidney has a certain blood pressure at which it functions at its best, that is to say, at which it can produce a high enough urinary output to control blood volume. However, if excess salt is taken in, then the *blood pressure will have to remain high* in order to maintain the normal level of blood volume. This is so because a higher urinary output will be needed by the kidneys to control the extra fluid.

Q. *Do Some Countries Have More Hypertension Than Others?*

In countries such as Japan, where there is an extraordinary high level of salt in the common diet, there is also a significantly high rate of hypertension. The reason for this, I think, is self-evident and more and more physicians and researchers seem to agree: salt stresses the mechanism that controls the fluid and blood levels, and high blood pressure comes about as a direct response to the additional salt load.

Dr. Lewis K. Dahl of the Brookhaven National Laboratories has been responsible for some of the most intriguing research about salt and hypertension. Dahl, wishing to explore why hypertension is so rare among primitive peoples with low sodium diets, investigated the correlation between salt consumption and hypertension. As reported in *Preventive Medicine*, Dahl conducted a pilot study at Brookhaven in which 1,346 adults were classified according to their intake of salt. There were three categories: low intake (never adds salt to food);

average intake (adds salt after tasting if insufficiently salty); and high intake (customarily adds salt before tasting).

Of the adults in the study, 105 were found to be hypertensive. Of all the subjects who had a *low intake* of salt, one was hypertensive. The high intake group had 61 cases of hypertension. Clearly, the tide is turning in hypertension research. Until recently, it has only been through testing laboratory animals that hard data on salt's role in hypertension has been available; now we are learning with more and more certainty that salt has the same destructive effect on human beings.

Kidney damage can, in and of itself, cause hypertension. A narrowing of one of the arteries leading to the kidney, for example, can cause the kidney to "perceive" that the blood pressure is low and to respond by secreting renin, which, of course, raises the blood pressure. In a case in which there is decreased blood flow in the kidneys but not in the rest of the body, the kidneys will continue to signal for the release of more and more aldosterone to raise the blood pressure. In other words, the blood pressure is essentially normal throughout the body but because of a malfunction of the artery of the kidney, the kidney behaves as if there were a systemwide blood shortage. Such a situation is extremely hazardous. The damage to the kidney will make it unlikely that the kidney's perception of its blood level will improve, and in the meantime the rest of the body is being flooded with blood and blood pressure is driven higher and higher. This is a kind of renal hypertension. Any sort of kidney damage will cause a drop in the kidney's ability to handle salt and it can no longer function constructively. It may turn out that a surgical correction of the kidney—or even its removal—will prove necessary.

It cannot be emphasized too strongly: The underlying

factor in all hypertension is some degree of kidney malfunction which does not allow the normal excretion of sodium. Clearly, in such cases the worst thing that you can do for your kidneys is to give them more and more sodium to process. In such a way you only emphasize and worsen the weakness. And, of course, it is not only the kidney that suffers from the excess salt but your entire body.

In a nation in which millions and millions habitually use excess salt in their daily diets, in a nation in which cheap, otherwise unpalatable foods are "jazzed up" with treacherous amounts of salt, in a nation in which the mania for the salty taste is really a kind of quiet drug addiction, it is no wonder that hypertension is so terribly widespread. Though the vast numbers of people currently suffering from this disease put it at plague proportions, there are, sad to say, still millions who have no idea of what hypertension is or whether they are suffering from it. Doctors tell us that one of the major problems in battling hypertension is that, at certain stages, it is often asymptomatic—that is, those who have it are unaware of it.

However, just because it is asymptomatic doesn't mean that hypertension isn't doing vital damage, even in its earliest stages. Too many people, doctors report, don't show up for a check of their blood pressure until hypertension has already done irreversible damage to their kidneys or their cardiovascular system. In other cases, patients *do* have symptoms of hypertension— headaches, dizziness, tingling in the limbs, visual disturbances—but they are so unaware of hypertension that they write the symptoms off to something else and treat it with aspiring or some other relatively ineffective over-the-counter remedy. And, all too frequently, as they do this they continue to ingest deadly amounts of salt.

The tragedy of our national inattention to this disease of hypertension is that the earlier it is caught the better the chances are for a complete reversal. Before vital organ damage is suffered it is possible to do yourself immeasurable good by simply altering your diet and kicking your addiction to salt. Of course, this is not all there is to it and that's why, when you're talking about something as critical as hypertension, it is always a good idea to consult a physician. But remember: When hypertension goes into its malignant phase—the phase of organ damage—and if it continues to be untreated, the chances are that you'll be dead within a year.

Q. *What Is The Medical Establishment Doing About It?*

The test for hypertension is, of course, simple, quick, and one hundred percent painless. In many areas throughout the country there are mobile hypertension units which screen communities for hypertension and which will give you a personal reading of your blood pressure without any charge. Today, the American Dental Association is urging dentists to give their patients blood pressure tests. Clearly, the alarm over the problem of hypertension is spreading and we can only be grateful for this. However, the fact remains that still millions of Americans are hypertensive without knowing it and that we as a nation are still told very little about the dangers of this disease. The national addiction to salt continues unabated and in far too many cases —in most cases, perhaps—physicians will not tell their patients about the perils of salt until salt consumption has already caused damage—damage which is, in many cases, irreversible.

How much simpler it would be if we knew from the beginning that salt was a potentially deadly drug.

It is crucial that people understand that they may be heading toward hypertension or even suffering from it already and not fully know it. As I studied the role of salt in disease I spent a great deal of my time visiting the hypertension clinics of New York City and speaking with doctors there. I also had an opportunity to speak to some patients and learn just how varied and unexpected the advent of hypertension can be. To give you an idea of the range of hypertension I'd like to give two brief case histories of patients I saw as I visited the hypertension and renal clinics of the city. In these cases, salt was a major contributing factor in the attack. Even in the case of prior structural damage to the kidney, the outbreak of the disease was, finally, directly attributable to the individual's consumption of excess salt. In both of these case histories the names of the patients have been changed. Besides consuming excessive salt, these two individuals have one other thing in common; neither one of them suspected that they suffered from hypertension.

Case History: Essential Hypertension

Penny Johnson was a thirty-three-year-old black stenographer. Since age twenty-two, she had suffered from severe headaches in the back of her head. She paid little attention to these and consumed large amounts of aspirin. One day, while shopping in a neighborhood department store, she consented to have her blood pressure taken by a nurse from a hypertension screening unit which was operating in the store for the day. The smiling nurse's face fell when she took Penny's reading. She told her that her reading was "extremely high" and to see her family physician immediately. Since Penny did not have a personal physician, she went to the emergency room of a nearby hospital.

She was admitted immediately; her blood pressure reading was 220/130. Examination of her eyes showed long-standing damage with some small hemorrhages of the retinal vessels—a cardinal symptom of hypertension. Her electrocardiogram showed an enlarged heart and there was protein in her urine.

Later she recalled that her grandfather had died of a stroke and that her cousin had died of kidney disease at age twenty-five. When asked about her dietary habits, it was revealed that Penny had a moderately high intake of salt—about ten grams per day (about four teaspoons).

Penny was placed on a diuretic pill to reduce the amount of fluids in her body, and strong anti-hypertensive medication. Penny will need close supervision for the rest of her life, and she will always be on medication. Her physician placed her on a low sodium diet.

Penny's story is very common. She had only mild headaches as a symptom, but when she was tested, her blood pressure was extremely high. The elevated pressure had already caused moderate damage to her heart and kidneys which will never be reversed.

Case History: Renovascular Hypertension

Barry Kaufman was a twenty-three-year-old college student from a white, middle-class background. During a routine physical examination for college, Barry developed a sudden headache. When his blood pressure was taken, it was 170/110. Checking his history, the physician found no reasons for the high blood pressure. Barry was not a diabetic, there was no family history of hypertension, and Barry did not consume what the doctor considered an abnormally high amount of salty food. Physical examination failed to reveal any telltale mark of a cause for hypertension. However, during the exam, the doctor did hear a sharp, high-pitched sound that mimicked the heart beat. This is known as a *bruit* and occurs because blood is trying to squeeze through an obstructed vessel. It was on the strength of this that he ordered an IVP (Intravenous Polygram) which revealed a smaller-than-normal right kidney due to the obstructed blood flow. From this picture, Barry was diagnosed as having renovascular hypertension.

Renovascular hypertension is the most common cause of secondary hypertension in the country. It is caused by an obstruction or stenosis in the renal artery, the vessel which takes oxygen-rich blood to the kidney. The obstruction caused a reduced blood flow to the affected right kidney thus causing it to remain small and undernourished. This type of hypertension is the second most common after essential hypertension. It is particularly prevalent in those under thirty.

Barry underwent surgical repair of the renal artery so that there was no longer a stricture in it and normal blood flow was restored to the kidney. After surgical repair, Barry's blood pressure was back at 120/80 and he did not need medication.

Renal vascular hypertension is very often reversible if caught in time before there is permanent kidney damage. In some cases it is possible to treat the patient with drugs and not operate.

Because his disease was caught in time, Barry will lead a normal life and should expect no further damage.

In both of these cases, the hypertension suffered by the patients could have been eliminated (or, at the least, greatly lessened) had the individuals made certain that they avoided salt in their diet. This would have meant never adding extra salt to their food, for the amount of sodium their systems could safely handle would have been easily obtainable from just normal eating. Barry and Penny had no inclination to pay attention to their sodium consumption because they never dreamed that hypertension would be a factor in their lives. They were wrong and you could be too.

Q. What About Contraceptive Pills?

As a postscript to our discussion of hypertension I'd like to mention Nancy, another patient I met while I journeyed through New York's hypertension clinics. Nancy, like so many others, was unaware of her condition until she went to an ob-gyn clinic for a routine checkup. Nancy had been taking contraceptive pills for three years and this was her first checkup in that time. (A dangerous mistake.)

When Nancy's gynecologist took her blood pressure reading she found that Nancy had a reading of 180/110 which was definitely abnormal. However, the doctor did not want to let this pass when she discovered that both of Nancy's parents had hypertension and that Nancy herself had a history of cyclic fluid retention during her menstrual period. Clearly, Nancy was heading toward serious hypertension trouble unless some strong actions were immediately taken. The gynecologist prescribed that Nancy stop taking the pill and before long Nancy's blood pressure was normal.

Hypertension has been found to be a *common* side effect of the contraceptive pill and today many thoughtful doctors will not give it to women with a history of hypertension in their family or with a personal history of cyclic edema or kidney disease. Even the remaining defenders of the contraceptive pill allow that it will raise a woman's blood pressure to some degree though not always above the normal range. There is still controversy over why this occurs.

Significantly, I think, the major components of oral contraceptives are combinations of estrogen and progesterone, both of which have a decided sodium-retaining effect in the body. It is argued that for many women the real damage from hypertension may take some twenty years to develop. The same timetable may be in effect, for that matter, in the consumption of excess salt, just as, in terms of the risk of cancer, it may take years of cigarette smoke or pounds of bacon in order for the damage to be noticeable. The timetables for disease are always tricky and the course of many ailments defy prediction. What we do know, however, is that the contraceptive pill causes hypertension. That may not be the whole story but, when it comes to protecting health, I think it is enough. There are, I believe, too many workable alternatives to the contraceptive pill to risk the murderous effects of hypertension.

CHAPTER SEVEN

Salt & Losing Weight

Our salty diets are a major cause of weight gain. Salt causes a condition known as *edema* which, aside from destroying your health in many other ways, adds pounds and pounds to your body, in the form of stored-up and useless fluid which literally dams up your tissues.

Obviously nothing is as important as being in good health and feeling truly well. Compared with the horrors of hypertension, kidney failure, and disorienting swings of mood, the question of looking slim does not seem so terribly important. But as anyone who has had a weight problem can tell you, obesity is in itself a kind of endless nightmare. The dangers of being overweight are, as most of us know, numerous. Carrying a lot of excessive weight can fatally overtax your heart. Aside from the very real health hazards of being overweight there is, especially in our culture, a great psychological price to be paid for obesity.

The United States is today the worldwide leader in diet spas, diet pills, diet exercises, diet doctors, diet books, diet cruises, diet encounter sessions, diet clubs, diet hypnotists. Perhaps the day is not far off when some shrewd real estate developer will build a Diet Town where the weight-conscious can live together in an environment of mutual abstinence and support. Once doctors who concentrated on weight reduction diets were thought to be practicing a species of medicine less serious than others, but now weight reduction is recognized as essential for the health of many millions and these doctors have been in the forefront of the new techniques developed to help people lose pounds.

Inevitably, there have been diet fads, and some of them have been shown to be more than a little dangerous. I must admit that I'm no stranger to some of these so-called miracle diets. Until I learned the secret of living salt-free I tried a number of ways to quickly shed a few pounds. I tried any number of crackpot schemes to get rid of the bloating I was plagued with before my menstrual period, and the beginning of summer would usually find me on one crash diet or another so that I could lose a few quick pounds before going to the beach. I realized that my weight problem was a minor one but until I kicked my salt addiction it was a nagging and a constant one as well.

Let's be clear about one thing before we go any further. If you're many, many pounds overweight you've got to face the fact that most of the excess weight is being carried in the form of fat and not fluid. While salt and fluid reduction will *definitely* make a big and a fast difference in your appearance and your weight, curing yourself of edema will not necessarily totally cure your weight problem.

Edema is a condition that comes about through a combination of high salt intake and high fluid intake coupled with an underlying inability to excrete the excess salt and fluid. We have already seen how this fluid retention can destroy your health. Now we must see how it spoils your appearance.

Q. *Where Does Salt Cause Us To Gain Weight?*

Stored water can dam up in the tissues around the stomach and add its own extra roll of soft, spongy weight. It can bloat your ankles so that no bones are visible and even your socks and shoes feel tight. Have you ever had to take off your rings because your fingers were swollen? That's what used to happen to me, especially during the summer months when my salt and liquid diet both increased.

The weight I gained during the summer months literally scared me. I had no idea that salt was the culprit and that as soon as I could get that crystal monkey off my back my problem would disappear. One particularly bad summer saw me refuse to go to the beach house I was renting with some friends. There was no way I was going to parade around in my French string bikini with those bulges in my figure. Sitting down I looked like a buddha. I went on one of my diets but the problem only got worse. My ankles were bulging and a favorite ring of mine got so tight that it hurt—and I couldn't even pull it off! I could almost *see* the water bulging through my skin and some of my joints looked yellowish and stiff. On Saturdays my fingers got especially bad and I finally realized that that was so because on Saturdays I went shopping and to the museums. For these pastimes I generally kept my hands at my sides where fluid could easily accumulate. What a dreadful summer that was! It was not the first time that such a thing had happened to me, but it was the worst. I cringe in sympathy with the thousands of others who suffer from this same weight-induced self-revulsion.

It would have been so easy to avoid. And, of course, it only got worse during my premenstrual cycle. Strange now when I think of it that I never linked the two. But that was before I learned how salt intake undermined the body. Maybe if someone had told me that I could cure my condition by cutting out—or even cutting down on—salt I could have saved myself a lot of misery. But, then again, perhaps I wouldn't have believed them. It may have sounded just too simple and too good to be true.

Q. *How Do We Avoid Edema?*

I had to watch my step. First of all, I had to make certain that my salt intake during the hot months was extremely moderate. (When I think of the people who respond to the heat with a salt tablet. . . .) Secondly, I made sure that I didn't quench my thirst with the sodium-preserved diet fruit punches I was so fond of. I like to drink when it's hot but I stick to potassium-rich liquids, especially fresh fruit juices. Incidentally, it's crucial that you stick to the *fresh* juices. Certain of the processed ones have sodium benzoate in them—and that's just what you want to avoid. Also, make sure that what you're drinking is juice, not just a fruit *drink*. These fruit drinks need contain only a minute percentage of real juice. The worst offenders are those concoctions called, for instance, orange *flavored* drink. These need not—and so, probably will not—contain any real fruit whatsoever. They may be entirely the product of some flavoring wizard who is working to increase his profit margin. To begin with, such liquids will not supply you with the precious potassium you need and secondly they generally contain a great deal of sodium.

I read of one patient who actually died from drinking one of these flavored drinks. She was on a strict, low sodium diet but some of the hospital staff apparently didn't know much about the ingredients of commercial foods. The woman drank her cup of flavored chemicals and within two weeks died of congestive heart failure. Had she been drinking a fresh fruit juice her diuretics would not have been competing with a fluid-retaining sodium compound, and the fluids would not have built up to deadly levels in her body. The hospital staff later isolated the killer: it was the sodium benzoate the fruit drink company had dumped into their product as a *preservative*. (Obviously, what was being preserved was the shelf-life of the product but not human life.)

Remember: Any sort of sodium compound such as sodium benzoate or sodium ascorbate causes the same problems as sodium chloride or table salt. It not only *pays* to read the labels of all foods—your life may very well depend on it.

Q. *What Is Anasarca?*

During my study of salt I had the occasion to visit a woman I will call Claudia, who was in the metabolic ward of a psychiatric hospital. Claudia was some twenty pounds overweight and often gained another fifteen pounds from edema before her period. This type of severe, generalized edema is called *anasarca*. Twenty-four hours before her period she weighed 155 pounds, by the end of her third day she was at about 145, and on the first day of her new cycle she was back to 140. This severe and debilitating shift in weight proved to be a great hindrance in her social life. She could only fit into one baggy dress and one pair of loose trousers for a third of each month. Both costumes struck her as deeply unattractive and she suffered from acute shame and embarrassment about her body. I don't think Claudia's psychiatric problems were wholly or even directly salt-induced, but there is no question that her unhappiness was profoundly worsened by the physical changes she went through, which *were* salt-induced. On her worst edemic days Claudia felt that her body was so large that she actually filled a room. She often feared that she was overpowering people by her sheer bulk.

We all understand how our image of our body determines so much of how we feel. Our view of our self-worth and our expectations for how the world will receive us are both intimately connected with how we see ourselves physically. Disturbed body image is a cardinal feature in many psychiatric disorders, particularly schizophrenia. If a person is on psychologically thin ice to begin with it can prove catastrophic to be forced to endure a radically altered body some ten days of every month. Claudia became so obsessed with the image of her bloated body that she began to avoid going to certain places where she feared she would not fit in. I don't mean "fit in" in the social sense but quite literally. For example, she wouldn't go near crowded buses or subways and there were many doorways she was afraid to walk through because she might get stuck in them.

More women suffer from cyclic weight gain than do men. However, it most definitely strikes men as well. One young man I spoke with at a New York hypertension clinic described the incredible distress and feeling of disorientation that accompanied his edemic weight gain.

"I feel like I'm a different person," he said. "I look down at my hands or I see my face in the mirror and I'm so fat that I don't recognize myself. I used to be the thin guy. I look at these fingers. Look at them! They don't belong to me. They're some fat guy's fingers."

(This man was subsequently put on a strict low sodium diet after being given diuretics to wash out all of the salt his body had hoarded. The result? His edemic cycle was finally broken.)

Q. *What about Salt and Pregnancy?*

Quite recently I spoke with a close friend who had just given birth to her first child. Several days before the delivery, she was struck by a sudden craving for, of all things, ham and eggs. She decided to satisfy the craving but she didn't feel like messing up the kitchen her husband had cleared up before he went to work. So she put on her coat and walked a couple of blocks to a local diner where she ordered her ham and eggs. Since the beginning of her pregnancy she had been instructed to cut back on her salt intake, but she decided that this one time it wouldn't hurt to have a little.

Within minutes her legs began to feel terrifyingly heavy and when she looked down at her ankles they were enormous. It had taken no time at all for her bodily fluids to collect and swell her up. Then, hot on the heels of this first most unpleasant discovery, she experienced new symptoms. She heard ringing in her ears and she was dizzy, as if she were perched on the tallest branch of a tree. In other words, her blood pressure had shot straight up. Unable even to leave the diner on her own, she was helped into a taxi by a waitress.

"I could have just as easily jumped over the moon as walked home," my friend said later. As soon as she was home she got in touch with her doctor. It was a good thing for her that her doctor was aware of the havoc that salt can cause in the body, particularly in pregnant women. He wisely asked her what she'd eaten and was able to come up with a quick, simple, and direct remedy —without any drugs. She was told not to eat any more foods containing sodium. Within twenty-four hours she was back to normal. The fluids were gone and a few days after that she was off to the delivery room where she gave birth to a beautiful baby girl.

Q. *What Are Diuretics?*

In nearly all cases of people suffering from salt overload, a cure can be effected by drastically reducing the amount of salt in the diet. However, some people will want a little "outside help" and for others simple salt reduction will not do the job quickly or thoroughly enough. Diuretics are drugs designed to rid the body of excessive fluid and they are the quickest way to cure edema. They are, however, powerful drugs and should *never* be taken without seeking a doctor's guidance as they themselves can damage the kidney unless used with restraint. There are several types of diuretic and a doctor will prescribe one depending on the exact nature of your condition.

Some diuretics will cause large amounts of your body's precious potassium supply to be lost. This type of diuretic works by decreasing the reabsorption of sodium by the kidneys, which also decreases water reabsorption and causes the patient to eliminate a lot of salt and water through urination. Along with this, you should be warned, are also lost *any water soluble nutrients*. Many dieters foolishly seize upon a diuretic and are understandably delighted with the rapid loss of five or so pounds. But without a real change in eating habits—without, that is, a breaking of the vicious salt and fluid cycle—this weight will soon reappear. "Water weight" can usually be lost much more safely and effectively by restricting salt.

Basically I think we should forget about diuretics unless they are recommended by a doctor. And even then their use should be questioned. Diuretics affect the endocrine system, one of the body's most delicate and complex. If used over a long period of time, diuretics lose their effectiveness and, at the same time, can leave you with a severe and dangerous potassium deficiency. Potassium loss can cause such diverse ailments as fatigue, muscular weakness, and possibly even depression. The loss of potassium may, many doctors suggest, harm the kidneys. (In severe cases, potassium deficiency can even bring on heart failure.) And now there are reports linking the overuse of diuretics with the development of diabetes.

The overuse of diuretics—the widespread need to decrease the amount of fluid our body retains—is another manifestation of salt's deadly role in our lives today.

CHAPTER EIGHT

Salt & Power

As my understanding of salt became more detailed I asked myself this question: If salt is so bad for us, why is it so common? As we all know, it would be hard to go through a day without coming across salt. It is everywhere. (Later on, we'll take a look at what is, to me, the most upsetting aspect of the salt problem: how we are manipulated and disregarded by businesses whose profit is based on our craving for salt.) Can it be so common and so dangerous at the same time? And if it's so bad, why do we crave it?

Of course, salt is not the only so-called common substance that is potentially destructive to humans. Many of the natural fruits of the earth turn out to be dangerous. Coffee, tobacco, opium, to name just three. The fact that a substance occurs in nature is no guarantee that it is good for us. And an alcoholic's pathetic craving for just one more drink is all the proof we need that our appetites are not infallible guides to good health. In beginning the struggle to protect ourselves from the potential ravages of salt we must begin by admitting that our salty cravings are not leading us in the proper direction. (Toward this end, I have included a "salt-enders" program as well as some of my favorite saltless recipes.)

Part of our current susceptibility to disease through excess salt is, it seems, a kind of evolutionary accident. Salt played an important part in the diet of our animal forerunners, beginning with animal life's emergence from the salty sea. Then the earliest humans came along. They gathered the fruits and vegetables of the earth and, with great effort, ate whatever animals could be felled with the primitive weapons then available. It is crucial to our understanding of how salt affects us today to understand that that primitive diet, eaten in times of great and constant physical exertion, was extremely potassium-rich. The body excreted enormous quantities of potassium and held onto each precious milligram of sodium—just as our bodies continue to do today, even though our diets are now sodium-rich and potassium-poor.

In other words, our systems were designed to process food that was high in potassium and low in sodium, to quickly get rid of potassium and hoard the sodium, and that design continues today, though it is no longer useful or even safe. In a very real sense, those who suffer from sodium overload and fluid retention are victims of an evolutionary lag. It is not at all inconceivable that one day the body will readjust itself to modern realities and learn how to better dispose of sodium while conserving potassium. But even if this isn't an evolutionary fairy tale it won't do *us* any good. We have to cope with the bodies we have!

As a salt craver, I did my best (or worst) to hold onto my habit. When I learned that my prehistoric ancestors craved salt and thrived on it I thought I had found a rationale to continue my life of salt abuse. But I have learned that our former need and use of sodium has little relevance to the facts of life today. What if humans once had wings? That would hardly be any cause for us to leap out of windows whenever we felt like it. The fact is that whatever the price our ancestors paid for their salt consumption, we alive today are quite literally risking our lives through salt abuse.

Q. *Did Our Ancient Ancestors Consume Large Amounts Of Salt?*

The question of how much salt our ancient ancestors

consumed is open to question. It could be that they managed to eat very little of it—thus increasing its mythological weight. In a fascinating paper in the medical journal *Circulation*, Dr. Edward D. Freis of the Veteran's Administration Hospital and Georgetown University School of Medicine, sheds some intriguing light on salt consumption and hypertension. As Dr. Freis points out, "hypertension is not found in unacculturated societies nor does blood pressure rise with age." After discussing a few of the possible explanations for this proven fact and dismissing each in turn, Dr. Freis concludes that "studies suggest that it is the lack of salt in the diet which accounts for the virtual absence of hypertension in unacculturated peoples."

An interesting proof of this explosive hypothesis is the work done by F. W. Lowenstein on blood pressure in the Amazon basin. In studying a tribe called the Mundurucus, Lowenstein showed that after their "conversion" to Christianity the missionaries also introduced the tribe to table salt. Soon, "although still living under relatively primitive conditions, the members of this tribe showed a rise of blood pressure with age and some had hypertension." There are other medical-anthropological studies that point to the same fact:With salt comes hypertension, even in so-called primitive peoples. L.D. Page, writing in *Circulation*, proved this in regard to the Solomon Islanders. A.G. Shaper writing in the *British Medical Journal* observed that among the Ugandan nomads, those tribes who had lived without the scourge of high blood pressure and/or hypertension were those with the lowest salt intake.

In reviewing the literature on salt and hypertension, Dr. Freis concludes that "when salt is not added to the diet hypertension is low or absent. In almost every recent epidemiological survey of unacculturated peoples, the importance of salt has been emphasized as the leading possibility for determining the presence or absence of hypertension."

Q. *Why Do So Many Of Us Crave Salt?*

It appears that humans developed their craving for salt quite accidentally. Just as the primitive peoples of today who are introduced to table salt soon develop a kind of addiction to it, so did humanity in general saddle itself with the taste for salt more or less by accident. That is, our original use of salt and our need to salt our foods heavily did not come into being because we just naturally craved the salty taste. In his brief but useful book on salt, Maurice Hansen speculates that one day, some 15,000 years B.C., someone came upon some meat or fish that had accidentally been left in salt and he or she discovered that the food was free from decay. And so, it is speculated, salt or a mixture of salt and vinegar was, from that time onward, used as a most common preservative of food. (Clearly, the challenge of preserving foods was one of the most pressing our ancestors had to face.)

There is no question that salt is an effective food preservative. Hansen tells us of Egyptologists who came upon some salted fish which had been covered in clay. The curious archeologists were not quite so curious to taste the fish themselves—it was, after all, some 2,000 years old—but they fed it to some dogs who, they reported, ate it with great gusto.

Q. *Why Is Salt Considered "Holy" By Some People?*

We see now that many of the old religious dietary laws were, in a sense, precautions recommended to believers who wished to avoid food poisoning or infestation by food parasites. For example, the spiritual significance of the Moslem and Jewish prohibition of pork may be somewhat obscure, but the sound and practical value of that old law is abundantly clear. It is, after all, only recently that *anyone* could eat pork without running the risk of trichinosis, a painful and often deadly infestation. In this way we see that what was once called holy law was really just excellent, healthy advice.

This same mixture of the healthy and the spiritual can be clearly seen in tracing the history of salt. In the days of antiquity, for example, Jewish mothers, who knew already about how salt preserved foods, would rub salt onto the skin of newborn infants in an act of preservation and purification, as if the child's body and soul could be protected from the decay of time. The ancient Jewish tribes also offered salt to Jehovah at harvest time, in a sacrificial spirit. Salt is synonomous with holy food in the ancient Hebrew language.

The crossover between discovering that salt was beneficial in preserving meats and fish to believing that salt is therefore beneficial to humans in general—and in quantity—turns out to be crucial in the history of the deadly condiment. As the word spread throughout the world that salt was a food preservative and as rites and superstitions developed around salt, the once little-used substance became a kind of mania throughout the world —a mania that still exists in our own times. Egyptians salted the coffins of their honored dead. Desert Arabs sealed vows of undying loyalty with salt. The ancient Mesopotamians were so impressed by the purifying powers of salt that they considered it to be a holy substance. The Old Testament's Book of Numbers mentions "a covenant of salt," referring to salt's presence in sacrificial meals. (No one dared break a covenant made over salt.) In the Europe of the Renaissance, honored guests were allowed to sit "above the salt," that is, close to the saltcellar. One's closeness to the salt dish was a reliable indicator of one's status at court.

The use of and belief in salt spread throughout the world and this resulted in two things that are central in the biography of this silent killer. The first result was that more and more people developed the taste for salt, not only as it occurred naturally in some foods but salt poured extravagantly over their meals. This growing taste for salt fed on itself as it increased, for we have already seen that those with added salt in their diet tend to crave salt, and the more they add, the more they crave. And so as salt established itself practically, religiously, and then as a matter of craving, the inevitable next step was taken. It was turned into a business.

How it comes to be decided that some substances are more precious than others is often a matter of mystery. The early barter and trafficking in salt has an element of random obsessiveness, in certain respects, especially when we read of salt being traded for ivory or gold. But, on the whole, cornering the salt market made very sound—and, at times, very sinister—sense from the earliest times.

Q. *Why Did Salt Become So Precious?*

The commercialization of salt is, I think, the real turning point in this crystal killer's career. Today, as we'll see later on, salt's link with the profit motive is a major—if not *the* major—factor in the grueling sodium overload which debilitates so many of us today.

In some parts of the world salt was used as money. Roman soldiers were paid for their services with salt, and the word salary is derived from this practice. Though the Chinese were extracting salt from the sea and from their sodium-rich soil as early as 2,700 B.C., American fighter pilots in World War II carried salt in their survival kits when they flew over China. Why? Because in some provinces salt was still used as money.

In parts of Central Africa where there was no natural occurrence of salt the importance of salt grew to the proportions of mania, until it was as valuable as gold or even human life. Salt was carried in a six-month journey from the Red Sea. Timbuktu, then a major Saharan city, became as important as it did because it was on a major salt route. Again, the hangover of this obsession was enormously long. When the Italians raided Ethiopia during World War II, they acted like most conquering armies and proceeded to raid the Ethiopian government vaults. There they found—bars of salt, stockpiled as if they were bullion.

Salt was for hundreds and hundreds of years a great source of wealth for rulers and merchants the world over. As the population of the world increased and millions were trapped into the lowest economic classes, salt became a still more crucial part of the daily diet. The medieval peasant consumed as many as twenty grams of salt *per day*—mostly, it is speculated, to enliven the porridges and gruel that made up his or her diet. (Today most doctors will admit that persons on high protein diets obtain enough natural sodium without adding a crystal of store-bought table salt.)

Q. *How Did Salt Affect Politics?*

Not only did merchants get rich off the world salt traffic but governments profited from it as well, usually in the form of high and arbitrary salt taxes. In sixteenth century France, the raising of the salt tax triggered a peasant rebellion. Although it is generally believed that this long and bloody protest was caused by the injustice of the tax hike, historians are now beginning to understand that the peasants were also fighting out of an innate craving for salt, which the new taxes would make less available to them. While the rightness of the peasants' cause is clear it is also true that they risked life and limb to protect an addiction to salt that was completely unhealthy. The French salt tax—the "Gabelle"—was finally abolished after the French Revolution of 1789, but until then it was used sometimes greedily and at other times cruelly. Salt used for tanning leather, for example, was poisoned so that leather workers could not steal any to bring home for their meals.

England also had a salt tax, which caused the same uproar and the same excesses as did the French tax. In England, as in France, this salt tax created a black market in the condiment and there were salt runners and salt smugglers operating in much the same fashion as drug smugglers operate in our own time. (The analogy is deliberate; for salt has the addictive and destructive properties of a drug.) Salt smugglers, when apprehended, were given long imprisonments and were often tortured.

Neither the French nor the English ever forgot the profits that could be made from taxing the poor for their use of salt. When the French established their imperial domination of Viet Nam in this century they made certain that the Vietnamese people paid a tax for the use of salt. Similarly, the British colonialists taxed the Indian nation on salt. It was also deemed a criminal offense for Indians to take salt from the sea. In 1930, Ghandi, whose efforts were ultimately to liberate India, began a symbolic march to the sea. It was a three-week trek and along the way Ghandi picked up thousands of followers. When they finally reached the shore, Ghandi picked up a lump of salt from the beach and licked it. This seemingly harmless act packed tremendous symbolic wallop. Indians all around the coastline followed Ghandi's example the next day and the British, defending their supposed right to rule the people of India, arrested thousands for their violation of the salt laws. It is interesting to note that the British never did concede the Indians' rights to salt, and the salt tax was not repealed until the British finally left India.

Q. *How Did The American Salt Industry Begin?*

American Indians first mined salt and boiled it down from brine springs long before European settlers arrived. The tribes of Nevada and Arizona mined shallow salt veins and those of the East coast had access to underground brine springs.

It took the European settlers, however, to turn the manufacture of salt into real commerce. In 1635, Samuel Winslow set up a small plant to boil sea water on the coast of Massachusetts and this salt supply was instrumental in stimulating trade with Great Britain. Fish and trapped furs could now be preserved with salt and traded for British manufactured goods. The American salt industry took a quantum leap forward when the Indians led the newcomers to the brine rivers and underground salt springs.

The largest of the springs at the time were in Saltville—formerly Buffalo Lick—Virginia. This spring was discovered by a woman named Mary Draper Ingles who was being held prisoner by a local tribe. While she was with them, they showed her where the salt deposits were and how to prepare usable salt from the brine. It turned out to be a rather costly error; perhaps the Indians underestimated the salt mania and greed of the white settlers. Mary Draper Ingles eventually found her way back to her own people and she soon led a group of settlers to the salt springs, which they now claimed as their own. She also showed them the techniques for boiling up brine. By 1800 a deep shaft was sunk into the springs and this country's first commercial production of salt began. It was, as we shall see, to have a short life.

North of Saltville, a man named Comfort Tyler, an ex-colonel from George Washington's army, discovered brine wells with the help of Indians living near Lake Onondaga, New York. As seemed to be the rule when a settler "discovered" something, Colonel Tyler ended up owning these brine wells and he was soon a major manufacturer of American salt. He began the salt trade in New York, and it was to accommodate this enterprise that the Erie canal was planned and built.

Across the continent, in 1820, the people of San Francisco were producing their salt by the solar evaporation of sea water. This industry grew rapidly and by the latter part of the century there were some eighteen salt companies. These companies hired Chinese workers, many of whom, ironically enough, made their way in the New World by farming salt from the sea as their ancestors had done three thousand years before them.

Q. *Did Salt Affect The Course of American History?*

As in Europe, people's dependency on salt was used for often ruthless political and/or economic ends in this country. The manipulation of salt supplies was an important tactic in the Civil War, for example. The Confederacy did not produce enough salt to meet its needs and had traditionally traded for, or purchased the bulk of, it supply from Europe. In 1861, however, the Union navy blockaded southern ports and prevented the importation of salt. By 1862, the price of salt went from seventeen cents to thirty-five dollars per bushel. Predictably a salt black market sprung up and profiteers bought up prewar stores of salt and resold them at skyrocketing prices. Without salt the Confederacy was thrown into increasing disarray and the Union, sensing this, added still another significant blow. On December 21, 1864, the Union captured and destroyed the salt works at Saltville. The salt works that Mary Draper Ingles had led the settlers to was destroyed after only sixty-four years.

In a sense, tracing this history reveals the schizophrenic nature of salt. Clearly, its use as a preservative was important in the betterment of life. Without quick means of transportation or other ways of protecting foods from spoilage, salt was essential for our distant ancestors. Yet even this benefit had its bad effects, since, as we have seen, the practical use of salt soon caused humans to worship it as if it were supernatural and this sense of awe and preciousness has certainly been an ingredient in our subsequent obsession with salt. By the time salt became an economic and political weapon it was doing as much harm as good and now, in our own time, it is the harm which outweighs the good.

Salt, of course, is not alone in this category. The first users of opium and morphine had only the relief of human misery in mind. They had no idea that their drug would turn into a violent destroyer of millions of lives, and while there are still plenty of valuable uses for salt in our time, especially in isolated areas and places without electricity, the overall dependence we have on salt is, as so many doctors have noted, nothing short of an *addiction*. Why do we buy salted foods? Why do we keep salt shakers on our dinner tables? It is most definitely *not* because we need salt. It is a craving pure and simple. And to the extent that we cannot imagine enjoying food without salt, it is an addiction.

CHAPTER NINE

Salt Killers:
How Others Profit From Our
Addiction To Salt

Salt is big business. Today, in America and throughout the world, businesses create and increase their profits by playing on our addiction to salt. The network of salt profits is enormous and complex. It includes, naturally, the companies which manufacture and promote the use of common table salt. But, unfortunately, this attack on our health does not stop there.

If you are not aware of the dangers of excess sodium and are not watchful of what you eat and drink there is no telling how much of this potentially perilous element you will consume daily. We are literally inundated with it. Salt and other sodium compounds are dumped freely into our lives. It is even present in the water supplies of nearly all the American cities.

For example, if you live in Los Angeles, just by drinking a cup of tap water you have introduced some forty milligrams into your system, and if you happen to be a resident of Crandall, Texas a cup of municipal water will dose you with more than 400 milligrams of sodium! (See the list in this chapter for a more complete picture of sodium in the public water supply.) Of course, tap water is not the only or even the major culprit in the current epidemic of salt- and sodium-related diseases. It seems that whole segments of our food industry rest on a foundation of salt. That is, the food manufacturers first carefully foster a *taste* for salt in the public and then stay in business by meeting and furthering that salt dependency.

78

Q. *How Does The Food Industry Make Certain We Are "Salt Addicts?"*

The American food industry wastes no time in planting the seeds of salt addiction in its sadly uninformed customers. In recent years, we've heard a lot about the ridiculous cost and scanty nutrition of processed baby foods and I can only hope that this means that there will be a significant trend away from buying these products. It is disturbing to report—and it was shocking to discover—that most commercial baby foods are among the most nutrition-poor and sodium- and sugar-rich foods available in the supermarket. No matter how cynical or grim your view of business morality may be, it is still difficult to understand how the manufacturers of these infant foods can rationalize the contents of their product.

It is nothing short of outrageous that this practice is allowed to continue, for surely if the baby food industry cannot conduct its business with honor, then we have a right to be protected from it by our government. The terrifying fact is that these ready-made foods offer our children deeply inadequate nutrition, which can set up lifelong syndromes of poor health.

Q. *Is Salt Really Dangerous For Children?*

Baby foods are instrumental in setting the stage for the child's lifetime of salt addiction. In a study conducted by Dr. Lewis K. Dahl of Brookhaven National Laboratories, the salt content of thirty different jars of baby food was found to be grossly in excess of the meats and vegetables from which they were derived. (Shades of the Middle Ages! As we mentioned before, salt was crucial to the medieval peasant because it made otherwise tasteless mush palatable.) When Dr. Dahl fed these baby foods to a special strain of rats, genetically predisposed toward hypertension, the salt content was enough to trigger hypertension in them. Most or all of us who have had hypertension in the family are already targets for this disease at infancy.

Adding to the case against baby foods is Dr. Dahl's discovery that the young are the most vulnerable to the deadly effects of salt. Young rats given a high sodium diet at weaning time were more prone to develop hypertension than rats initiated on the diet at an older age. This is partly true because babies are even less capable of excreting excess salt than are adults and it builds up quickly, damaging the kidneys. This induced hypertension *remained* even after the diet was changed; the initial high sodium dosage was enough to seal their fates. It is, the study suggests, more than possible that feeding your child commercially prepared food will guarantee a lifetime of high blood pressure.   If there already is a history of high blood pressure in the family then particular care must be exercised.

Dr. Dahl tells us that, "The modern diet for infants may contain concentrations of sodium chloride so high as to result in average daily intakes comparable with the highest reported in man." This is, of course, utterly shocking. But what disturbs me nearly as much as Dr. Dahl's shattering findings is *when* his report was released. Dr. Dahl's experiments were first conducted as far back as 1957, and still there has been no forthright action, either on the part of the food industry or the Food and Drug Administration, to protect our young from this known danger in their foods.

Q. *How Common Is Hypertension?*

It is difficult to say with real accuracy how many Americans currently suffer from high blood pressure, partly because so many do without its being diagnosed.

However, most estimates put the figure at around 25,000,000. It is, with no exaggeration, at plague proportions. We have already proven that high blood pressure begets high blood pressure and so, if the chain is not broken, we can expect the number of hypertensive Americans to grow. It would not be accurate to blame this widespread affliction solely on salt, but even Dr. Dahl, writing with the restraint characteristic of a scientist, says, *"High intake of sodium chloride in infancy might play an important part in the propagation of hypertension in adults."* (American children are not alone in being victimized by a careless, profit-motivated diet. A recent study of French school children revealed that more than 10 percent of those tested were suffering from hypertension. This was directly correlated with the amount of salt in their diets.)

Q. *What Is The Purpose Of Adding Salt To Baby Foods?*

Clearly, there is no nutritional need for salt to be added to baby food. It is universally accepted that children would thrive on mother's milk, which is extremely low in sodium—five times lower than cow's milk even. So any argument that the salt is added because it is important for the infant's health is absolutely unwarranted and groundless.

So why do the manufacturers of baby food add salt to their product? Some say that the salt is dumped into these products to make them more appealing to the *adults* who buy them. (The same seems to be true in the manufacture and sale of dog and cat foods—the trick is to make it appeal to the owners who, after all, have the money. And so we have the moronic phenomenon of kitty burgers and bacon-flavored dog chow.) Courting the favor of salt-jaded adult palates, baby food businessmen try and make their product seem more like "real food" by riddling it with the same unnecessary salt that adults consume in their daily diets.

But this practice has a double benefit for those who profit from the sale of food. Not only do they stimulate sales by salting their product but they also introduce generation after generation to salt addiction, thus insuring its perpetuation. Just as psychologists report to us that children who are abused are extremely liable to become child abusers themselves (the "battered-child syndrome") so now it seems that babies who are loaded with salt are liable to load their children's diet with the same deadly ingredient.

Q. *How Can We Protect Our Children From Deadly Salt Addiction?*

I personally feel that breast feeding is the best of all possible worlds for both the child and the mother, nutritionally, immunologically, and psychologically. I urge the parents of older babies to prepare their own baby food. Now that blenders are available for not very much money there is no reason not to do this. Not only will you save a lot of money but you will be giving your child immeasurably more nutrition. And you will *know* that your child is eating soundly. Making your own baby food can help break the vicious cycle of salt overload. Your baby will get all the sodium he or she can use from natural sources such as meat or dairy products. You need never add a grain of salt to your baby's food. If your child never gets accustomed to having his food dosed with salt he or she will probably never ask for it.

But the disturbing fact is this: most children *are* started on their salt addiction in the earliest months and they are liable for the rest of their lives to have the addiction catered to and exploited. Think of how many foods would simply disappear from the marketplace if they were not loaded with salt. Can you imagine, for instance, a saltless potato chip? Take a walk through the "snack" section of any American supermarket and what you will find is low-grade starch jazzed up with killing doses of salt—for without salt they'd taste pretty much as what they are and few would be tempted to buy them.

In hundreds of instances, product appeal seems to be based wholly on playing into our conditioned, reflexive drive for salt. Buy a box of your favorite crackers and rub the hard crystals of salt off one and see what it tastes like! The manufacturers of the various potato chips, cheese curls, sesame snaps, potato sticks, taco chips, cocktail crackers would be doing a much less profitable business if the very real dangers of salt consumption were widely understood. Unfortunately, most people don't begin paying attention to their salt intake until they are seriously ailing from it and often, as in the case of my grandfather, not even then.

I realize that this is a free-market economy and our philosophy is "let the buyer beware." But isn't it time that we were at least warned, that we were told that salt is not just a flavoring but a potentially destructive drug? If the government will not step in and force manufacturers to stop salt poisoning the American consumer then we at least ought to have some notice similar to what we find on tobacco products: *Caution— Doctors have determined that salt is dangerous to your health.*

Q. *Do We Always Have To Be On Our Guard?*
Americans at leisure are constantly bombarded with an endless barrage of salt. Sometimes it is given away, such as the "generous" helpings of salted peanuts, chips, and pretzels some shrewd bar owners place out for their

customers. This, of course, costs the bar owner very little and it makes for nice public relations. But more importantly these salt-soaked tidbits also stimulate thirst and so, with their water retention system already working overtime because of the excess salt, the hapless consumer orders drink after drink. This of course increases the bar's profits and adds to the bloat, irritability and sometimes fatal dangers of fluid build-up.

At the movies we have a national habit of buying popcorn. It tastes salty even without our adding an extra grain, but as if the popcorn weren't deadly enough we then ask them to butter it, which adds still more salt, and then we give it a few extra shakes of salt before heading in to see the film. A large percentage of the popcorn consumers will soon be back, dry-mouthed, looking to purchase a cup of soda, which is a little like taking strychnine to counter the effects of arsenic.

Sports events are also great occasions for selling salt to the uninformed. It has always struck me that the potential dignity of our sports events is seriously undermined by the frankly *rotten* food that we allow to be sold during them. A child taken to witness our national pastime is surrounded by a nonstop army of vendors—some selling sugar, some selling alcohol, some selling pork scraps preserved with nitrates, and some selling salt in the form of salted peanuts, potato chips, pretzels and the like. It may be a real "treat" to bring a child to a sports stadium and let him cheer the team on, but you're not doing the child a favor if at the same time you feed him foods that undermine his health. Why not bring your own snacks? There are so many things that children positively *love* and that are good for them. There's no reason—and no excuse—to slowly poison the ones you love just because the salty foods are convenient or because the children claim to "like" them. Adults wouldn't dream of starting children off on most other addictions. Children are, for example, generally denied access to cigarettes and tranquilizer pills, and this same spirit of concern should govern our behavior when it comes to salt. Of course, as has been proven with tobacco, you'll have a better chance in protecting others from addiction if you yourself aren't an addict. The best guarantee of a salt-free child is a salt-free parent.

Q. *Are Only the Foods That Taste Salty Bad for You?*

The profit that is made from salt and other sodium compounds goes beyond the salt that we detect as a salty taste. So few of us actually know how much sodium we dose ourselves with daily. That steak we buy at the supermarket, for example, probably has sodium—in the form of sodium nitrite—in it even before we put a grain of salt on it. The sodium nitrite not only counts as sodium in your diet but, many physicians, biochemists, and medical researchers believe, can cause cancer as well.

Sodium nitrite is present in all processed meat. It is also sometimes put on older meat to make it look moist, fresh and red. There are indications that one day nitrites (and nitrates) will be banned from commercial meat processing, just as the government finally got around to legislating against the deadly red dye no. 2. But in the meantime we must be on the lookout for it. And we must also realize that it usually takes official circles a very long while to react to the clear and present dangers in our diet. Recent history has more than proved that.

It is argued that sodium nitrite is not harmful in itself. Former generations knew it as saltpeter and it has long been used to preserve and cure meat. The meat packing business argues through its official publications and its busy Washington lobbyists that sodium nitrite is no health hazard whatsoever, just as the aerosol spray people scoffed at the reported dangers to the ozone layer and the Thalidomide profiteers claimed their drug was perfectly safe. In the opinion of the meat packing industry, the nitrites are necessary to combat bacteria and, they say, to improve the taste of their product.

Both of these contentions are highly questionable, as we shall see. But even if they were true they would mean very little in light of the dangers of nitrite consumption. More and more scientists are now realizing that sodium nitrite can cause cancerous tumors—even if ingested in small amounts. In fact, there is now evidence that small amounts of nitrites ingested over a period of years is the *most* efficient way of producing a cancerous tumor.

Q. *How Dangerous Are These Nitrites?*

Sodium nitrite at levels higher than ten parts per *billion* constitutes a potential cancer hazard. Sodium nitrite is present in very high levels in hot dogs, bacon, ham and canned fish. In spite of the fact that no official tests to determine sodium nitrite levels in our meat were ever conducted, the FDA nevertheless allows two hundred parts per million in processed meats. How did they come up with this incredibly dangerous figure? It was simple. That was the average amount of sodium nitrite found in cured hams in 1925!

It is now agreed, even by nitrite's defenders, that animals feeding on high levels of the chemical will develop cancerous tumors. Nitrites combine with reactive amines found in meat, eggs, liquor, and even cigarette smoke to form substances capable of producing cancer. Shockingly, many prescription drugs also combine with sodium nitrite to form a carcinogenic compound. In much of our drinking water—especially in areas where synthetic fertilizers are in use—there is an incidence of nitrate salts which are converted into nitrite salts by bacteria in the intestinal tract.

Another way to ingest these dangerous compounds is in cooking. In frying bacon, for example, the nitrites vaporize in the heat and not only do you eat the sodium nitrite in your bacon but you inhale fully four times that amount while you're standing over the stove. The higher the temperature, the more readily do the poisonous compounds form, and the more strength they seem to have.

Q. *Why Do Manufacturers Put Sodium Nitrite In Our Food?*

Unlike sodium chloride—salt—sodium nitrite is not dumped into our foods because food manufacturers have fostered an addiction for the stuff. Basically, the potentially fatal compound is used so widely because it is a quick, inexpensive way of preserving some foods. Whatever the danger is seems much less important than increasing the profits of the food companies. But a certain dependence on the part of the consumer is also a factor in the widespread use of sodium nitrite. The nitrites give some meats the artificially red appearance that some people seem to enjoy. It's an extremely high price to pay just to see meat a fake bright red! In other cases, the nitrites give meat a "cured" taste that other people enjoy; again the taste is fake and the price is incredibly high.

Some businessmen claim that a product like bacon could not be sold without nitrites. Bacon, they argue, can't be processed under high temperatures or else the fat would melt. This is an interesting argument since it reveals how the bacon sellers depend on selling a product unconscionably high in fat. Yet even in the case of bacon the arguments for the "necessity" of nitrites is a false one. First of all, it would be infinitely better not to have any bacon whatsoever if the only way to eat it is to eat along with it cancer-causing chemicals. And, secondly, it has recently been demonstrated that even bacon can be cured and shipped without nitrites. There is now a product called Kennedy's Homogenized Bacon, made by an Iowa meat packer, which is completely nitrite-free. (To do this, he melts away most of the fat and forms the meat into patties which he homogenizes.)

Clearly and without question, there is no excuse for the use of nitrites in our food. Even if the evidence against these chemicals were *less* convincing, even if there were only a faint possibility that they may cause cancer, it would be reason to ban them. We are, in our time, in the midst of a cancer incidence that is tantamount to a plague. Many of these cancers, it is now generally believed, are environmental. That is, the factors that are killing so many millions are the air we breathe, the quality of our workplaces, and the foods we eat. That we can know this desperate fact and still allow nitrites in our food defies comprehension.

Q. *Are There Other Dangerous Uses of Sodium By-Products?*

Another way in which sodium is slipped into your diet is through monosodium glutamate, better known as MSG. It is perhaps most closely associated with Chinese food but it is also widely present in many canned, frozen and packaged foods. MSG is what is called a "taste enhancer." That is to say, the action of this compound somehow causes food to taste more attractive. It has, clearly, no nutritional value and it is added to foods only to make the consumer more eager to buy and eat them.

There is a condition known as "Chinese restaurant syndrome" that is wholly the result of MSG. People who are sensitive to this additive react with headaches, dizziness, stomach pains, and sometimes even fever. Others will develop edema from the sodium. And, of course, all of the other ailments deriving from excess sodium can be exacerbated by MSG. MSG in all of its forms is particularly dangerous for those who must be careful about their salt/sodium intake. Even if it is put in with a low sodium food such as stir-fried vegetables, the MSG defeats the whole purpose of eating such safe foods.

A trip through the supermarket will reveal the incredible prevalence of MSG—even bouillon cubes are full of it! It is even sold in the shaker, to be showered like salt onto our food. MSG actually tastes less salty than sodium chloride but it has three times the sodium content.

Another widely used preservative is sodium benzoate. It is often mixed in with salad dressings or relishes. Baking powder and baking soda used to leaven bread are also sodium compounds. Sodium sulfite is used to *bleach* fruits which are then colored to look more commercially attractive. Sodium ascorbate is a type of Vitamin C, but a few years ago this form of the vitamin was taken off the market because the government determined it was dangerous to unsuspecting consumers on a low sodium diet. Why, one wonders, was this one example of hidden sodium legislated against when all the others continue unregulated? It seems that the sodium ascorbate manufacturers were brought into line only because the vitamin industry is less powerful and influential than the meat packing industry and all of the other large businesses which continue to reap profits through salt and other sodium compounds.

Q. *How Do We Protect Ourselves From Hidden Sodium?*

As business goes on its merry, deadly way and government chooses to sit on the sidelines, we must learn to protect ourselves from the hazards in our foods. The first step, of course, is to read all labels carefully, word for word. You will be shocked at all of the salt and sodium-based preservatives and "taste enhancers" you will find. You will find them in everything from peanut butter to gefilte fish. All such products *must* be dropped permanently from your shopping lists.

We must allow our taste buds to get over the years of salted trauma so we can appreciate the natural tastes of foods. For instance, when I really *thought* about and experienced the taste of frozen chicken and noodles in cream sauce I realized that it tasted as salty as brine and had an unnatural chemical edge that didn't really camouflage the bland, cooked-out taste or the insipid texture of the sauce or the chewy chicken cubes. True, I have given up the mindless convenience of much prepared food, but I've found that now my meals are not only healthier but are infinitely more tasty—if I do say so myself! I realize that some of us don't have the time or inclination to expend much effort in the kitchen on a nightly basis and so in the next section I give a few of my favorite saltless recipes that are also very fast and simple.

Another way of saving time and avoiding salt is to prepare saltless dishes and then freeze them. If you buy commercially frozen or canned foods, however, you'd better give them a good washing before cooking or else you'll be eating a lot of salt. Most plain frozen vegetables don't have salt, except for lima beans and peas, which often do. When it comes to simple meals, however, one thing you must give up forever is canned soup. This product has an outrageously high sodium content. Clearly, a good rule and a reasonable way to protect yourself from continued salt addiction and the other ill effects of sodium consumption, is to stay away from refined and processed foods. It may seem like a lot of extra bother to begin with but after a while it will be automatic. And the benefits in continued good health or in the improvement of previously impaired health will be more than worth the small effort. Living salt free is a matter of feeling good.

Q. *Are "Water-Softeners" Harmful?*

Another hidden source of sodium in our lives is water softeners. We call water "hard" when it contains large amounts of minerals, usually calcium and magnesium. Hard water is considered to be a nuisance by many, who say that it is harder to get dishes and even themselves completely clean washing in hard water. Hard water can also leave deposits in pipes and dirty rings in the bathtub. And so to combat this many buy commercially produced, mechanical water softeners.

These water softeners, sold with no warning to the consumer, work by replacing the hard minerals with sodium chloride. An ion exchange water softener consists of two tanks; one has a filter bed made of synthetic resin, and the other stores sodium chloride which is dissolved in small amounts to form brine. The hard water is pumped into the resin filter where it exchanges its minerals for the sodium ions. In this way you can bet on getting from 200-500 extra milligrams of sodium every day from your water!

I think these water softeners ought to be sent back to the companies which make them or should be left to gather dust on the shelves. But if you are so annoyed by your hard water that you can't stand it then please only install your softener on the hot water tap and never use the softened water for cooking or drinking.

Hard water may not give you the mountains of soap suds that you see on TV, but many studies have demonstrated that hard water is good for your heart. Precisely why is not yet fully understood, but studies show a correspondence between hard water and lower rates of heart disease. Dr. Henry Schroeder of Dartmouth tells us that cities such as New York where the water is exceptionally soft have twice the death rates from coronaries as hard-water cities. Dr. Schroeder believes that this fact may be explained by soft water's tendency to pick up poisonous substances from pipes—particularly cadmium, which is found in iron-galvanized pipes. In soft-water areas, it's a good idea to let your water run about ten minutes in the morning to rid it of the build-up of metals such as cadmium and lead. That way you won't be so likely to drink these potential poisons with your morning tea.

One town in the south installed a water softener for its municipal water supply and within one year the number of coronaries had risen significantly. Shocked and dismayed, the town voted to return to hard water, realizing it was more important to live than to be up to your elbows in lemon-scented soap bubbles. In my tours of hypertension clinics I talked to one man whose doctor told him to remove his mechanical water softener from his water supply. Before that he had been diagnosed as having essential hypertension but as soon as the softener was gone his blood pressure returned to normal.

In some cities the municipal water supply is very high in sodium. It is best to check with whoever regulates such things in your town to make sure you're not consuming dangerous amounts of sodium with each glass of water. (Soda drinks, by the way, aside from everything else that is wrong with them, are made from sodium-rich water.) Here is a very partial list of some cities and the amount of sodium found in the water. Since now doctors believe that taking 500 milligrams of sodium is safe for daily use, you'll see how far along toward the number you go just by drinking the water. (This list was taken from *Cooking Without a Grain of Salt*, a valuable resource book written by Elma W. Bagg.)

PUBLIC WATER SUPPLY   MILLIGRAMS PER CUP

Public Water Supply	Milligrams Per Cup
Albany, N.Y.	0.4
Albuquerque, N.M.	12.0
Ann Arbor, Mich.	4.8
Atlanta, Ga.	0.4
Austin, Texas	7.2
Baltimore, Md.	0.7
Baton Rouge, La.	21.6
Biloxi, Miss.	55.2
Birmingham, Ala.	4.8
Boise, Idaho	4.8
Boston, Mass.	0.7
Charleston, S.C.	2.3
Charlottesville, Va.	0.4
Cheyenne, Wyo.	0.7
Chicago, Ill.	0.7
Cincinnati, Ohio	1.6
Cleveland, Ohio	2.3
Crandall, Texas	408.0
Dallas, Texas	7.2
Denver, Colo.	7.2
Des Moines, Iowa	2.3
Detroit, Mich.	0.7
Hartford, Conn.	0.4
Helena, Mont.	0.7
Houston, Texas	38.4
Huntington, W. Va.	7.2

PUBLIC WATER SUPPLY	MILLIGRAMS PER CUP
Jacksonville, Fla.	2.3
Kansas City, Kan.	9.4
Kansas City, Mo.	23.6
Lincoln, Neb.	7.2
Little Rock, Ark.	0.2
Los Angeles, Ca. : aqueduct	14.1
metropolitan source	40.1
river	12.0
Louisville, Ky.	4.8
Miami, Fla.	4.8
Memphis, Tenn.	4.8
New York, N.Y.	0.7
Omaha, Neb.	18.8
Phoenix, Arizona	25.9
Philadelphia, Pa.	4.8
Pittsburgh, Pa.	14.1
Providence, R.I.	0.4
St. Louis, Mo.	12.1
Washington, D.C.	0.7
Wilmington, Del.	1.8

In closing this chapter, let's just touch on three more sources of unwanted sodium. One is in such stomach remedies as Alka-Seltzer. Another is in some toothpastes which contain a great deal of sodium. (Be sure and rinse doubly and even triply well.) And the third is the common American postage stamp. The glue our government puts on our stamps is very high in sodium. If you're stamping a lot of letters it really makes sense to use a sponge.

CHAPTER TEN

Kicking/Saltless Cooking

All habits are difficult to break. It seems that the body likes nothing more than to repeat past actions, no matter how nonsensical or even dangerous they are. If your fingers are used to drumming on table tops you'll have to really fight with them to make them stop. If your lungs are used to sucking in polluted, carcinogenic cigarette smoke it will take an act of will to break them of it. The same habit-breaking process will be necessary when you kick the salt addiction—though there will be no "withdrawal." In point of fact, of all the addictions the human body is prone to—and isn't it amazing all of the things we can fall prey to?—salt is one of the quickest and simplest to handle.

In my case I have been known to literally pour salt onto my palm and lick it up, so fond was I of that salty taste. But after just a couple of months of kicking my salt addiction I found that there were some restaurants where the food was unbearably salty to my taste. What happened was that my taste receptors had quickly readjusted and had already begun to crave the natural, nonbriney taste of real, unsalted food. Even a "normal" amount of salt tasted astringent to me.

Q. Won't Food Taste "Boring" Without Salt?

There is, or so I've been told, an old Spanish proverb that says, "An egg without salt is like a kiss without a mustache." This is exactly the sort of thinking you will have to leave behind when you decide that the time has come to stop ruining your health with sodium chloride. As one who has been living salt free for a couple of years now, I think it's just incredible that people still

95

think that salt *improves* the taste of food. (We won't even *talk* about whether or not kisses are improved by mustaches!)

The simple, gastronomic fact is this: Salt does not improve the taste of food—it masks it. There *are* seasonings which do add resonance and subtlety to cooking and those are spices and herbs.

It is true, of course, that cooking the saltless way makes great sense from the standpoint of good health maintenance but, as I have now learned, it also makes great sense from the gourmet's point of view. I'm not now, nor will I ever be, cooking at the level of the world's great professional chefs, but there is no question that my meals have fantastically improved since I kicked my salt addiction and began expanding my understanding of spices and herbs. There's been a dual improvement. My food is no longer marred by the vulgar taste of salt and I have learned so much more about the varied and surprising uses of such seasonings as tarragon, basil, and ginger.

Of course one of the great benefits of living the salt free way—both in terms of health and in terms of enjoying food—is that I no longer eat the salt/sodium loaded products of the American mass-food industry. As I have tried to stress, the more processed foods you eat the more sodium you are eating. This is a simple and vital precaution, but it is also a gateway to better eating. Because another rule of thumb is: The less processed food you are eating the more you are likely to enjoy your meals.

I no longer buy or eat salted butter and believe me this is no sacrifice. The unsalted butter is infinitely better. (I've since learned that most of the best cooks insist on unsalted butter, not only for eating but for cooking.) Unsalted butter is fresher (it *has* to be) and it is sweeter. And it isn't artificially colored like so much of the "regular" butter is.

I used to do something that I thought was a clever cooking trick. When I was to boil something—say, vegetables—I added salt to the water. Now I've learned that not only does this add potentially dangerous sodium to my diet but that the salt actually draws the vitamins out of the vegetables and allows them to cook away.

Q. *How Do I Begin?*

When you decide to kick your addiction to salt you can easily do it "cold turkey," that is just *stop* salting your food. (Making certain, of course, that you are getting at least 500 milligrams of sodium in the course of your day.) If this sounds a little drastic let me assure you that it's not at all difficult. However you may want to take it slow.

You can begin by not salting your food before you at least *taste* it.

Then you can make sure that you put on as little salt as you can possibly manage. Try tasting your food after sprinkling on a couple of grains of salt.

Don't allow any salt-added foods into your kitchen.

Stock up on saltless snacks.

Reacquaint yourself with fresh fruits and vegetables.

Don't salt any food while it's cooking.

After a week of this you'll be more than ready for the important step of removing the salt shaker from your dining room. My advice is to throw it away—unless, of course, it's a silver heirloom, in which case you might want to show it to your grandchildren. But whatever you do with it, remove it from its former place of honor on your table. Remember it contains as much potential danger as a container of strong medicine. Once you understand that salt has many of the properties of a drug you'll be able to handle it safely and intelligently. The only use I have for the salt I buy now is to mop up spilled wine from my grandmother's damask table cloths and napkins.

Q. *What Should I Use Instead Of Salt?*

Your table needn't be bare without your salt shaker. Put out an attractive container of basil leaves or sesame seeds, or some spice combination of your own invention. If you enjoy seasoning your food as you're eating it there's no reason to give that up. All you want is to stop poisoning yourself with excess salt. You don't have to stop enjoying your meals.

Try growing fresh herbs in your kitchen. Most of them love to grow and will come up under all kinds of conditions. If you buy dried herbs buy them in small quantities so you'll be continually getting fresher supplies. Make sure to keep them in a fairly tight container.

If you want to add some wine or sherry to your food while it's cooking there's surely nothing wrong with that. I always keep on hand a jug of inexpensive wine, both a red and a white. For some dishes it adds a terrific jolt of flavor. Please avoid the so-called "cooking wines" sold in many supermarkets. Some of them contain forms of sodium. *Read the labels!*

With these points in mind, I'd like to share some of the salt-free recipes that I've been enjoying so much over the past couple of years. After each recipe I give the approximate amount of sodium a serving will contain.

Q. *What About The Low-Sodium Salts?*

There are now products available for people on low-sodium diets. In many cases, a low-sodium product will still give you more sodium than you need. However, I can only think that the existence of these products is a step in the right direction—even if only a small step. Some health food stores, or health food sections of supermarkets, sell low sodium mayonnaise and low sodium bread. Neither of these is a real substitute for the safety and goodness of making your own, but they may be used instead. There are also available a number of low-sodium artificial salts. I personally think that using a salt substitute is unnecessary when it's so simple and enjoyable to live salt free. However, you should know that these low-sodium salts are not sodium free. Most are a sodium-potassium mixture. When you decide to kick salt you had best ask your physician about cooking with these so-called "lite" salts. If he wants you to use one you should ask for a specific brand name. The compositions of these artificial salts vary from brand to brand.

Vegetables

Vegetables should always be steamed so that they retain their vitamins and minerals. They should be eaten crisp instead of the sorry rags that so many people pass off as vegetables.

To me nothing tastes better than crisp vegetables with a little oil, a squeeze of lemon juice, a clove of garlic and some pepper. Those of you on low calorie diets should fill yourself up on plenty of fresh vegetables.

Most vegetables are low in sodium and you can eat as much as you want of them. However, the following are quite high in sodium and you should check the tables to make sure they don't throw off your daily allowance: Artichokes, kale, beets, chard, beet greens, spinach, celery, white turnips, and dandelion greens.

high sodium

Baked Acorn Squash (2)

1 large acorn squash
2 tbls. sweet butter
4 tbls. of molasses

Cut the squash in half. Add a tbsp. of sweet butter and 2 tbsp. of molasses to each half. Bake in oven for 45 min. at 350°.

43.8 mg. sodium per squash

Sauteed Zucchini (4)

1 lb. zucchini
1 tbl. butter
½ tbl. olive oil
1 tbl. minced shallots
1 clove garlic, minced
pepper to taste

Cut the zucchini into small pieces. Blanch them by dropping them into boiling water for 1 minute. Saute the shallots and garlic in the butter and oil. Cook until transparent. Add the zucchini, cover and saute gently for 10-12 minutes, stirring occasionally so that the zucchini is lightly browned on all sides.

4 mg. sodium per serving

Sauce Vinaigrette (½ cup)

6 tbls. olive oil
2 tbls. wine vinegar
½ tsp. dry mustard
1 clove garlic, crushed
any herbs you wish to add

Shake or whisk the ingredients together. I prefer to let this mixture sit together overnight and strain out the herbs the next day. This is important if you are using dried herbs, because they can give the dressing a scratchy texture.

Negligible sodium

Curried Sauteed Eggplant (2)

1 small eggplant
4 tbls. olive oil
curry powder*
pepper
½ cup flour

Slice the eggplant into thin rounds. Dredge each piece in flour. Heat 1 tablespoon of olive oil and add as many pieces of eggplant as will fit in the frying pan. Sprinkle them with curry powder. Turn, saute until golden brown. You will have to add more oil to the pan from time to time so that the pieces don't stick.

10 mg. sodium (This varies with the size of the eggplant.)

* Some curry powders contain salt. You may use it unless you are allowed no sodium at all in your diet.

Stuffed Mushrooms (4)

12 large mushrooms
¼ cup melted butter
1 tbl. parsley
1 medium onion, minced
1 tbl. sherry
1 clove garlic
1 tbl. oil
4 tbls. wheat germ

Wash the mushrooms and cut off the stems. Chop the stems and add them to the onion, garlic and parsley. Add the sherry and saute this mixture gently in 1 tbl. of oil. Stuff the mushrooms with this mixture and put them underside up in a baking dish. Sprinkle them with wheat germ and pour the butter over the top. Bake for 15 minutes at 400°.

3 mg. sodium per serving

Green Beans Vinaigrette (4)

1 lb. green beans
¼ cup Garlicky vinaigrette sauce

Steam the beans until they are bright green and still crunchy. When ready, mix the beans and dressing with clean hands. This way the beans are coated evenly with dressing. Let them cool in the refrigerator or they will absorb the dressing and become soggy.

Negligible sodium

Rice Ideas

Put a large pinch of saffron into the cooking water. Saffron rice is especially good with chicken dishes.
Add butter and ½ tbl. curry powder to 2 cups of cooked rice. Stir in some freshly chopped tarragon, thyme or chives.

Baked Stuffed Potatoes (4)

4 potatoes
2 ounces hot milk
1 tbl. butter
chives or tarragon
pepper

Bake the potatoes until they are done. Split them and carefully scoop out the meat. Add the butter and mash the potatoes with a fork, gradually adding the hot milk, and continuing to mash until the mixture is smooth. Add chives or tarragon to taste. Put the mixture back into the potato skins and bake them at 400° for 10-15 minutes until the potatoes are lightly browned on top.
10 mg. sodium per serving.

Jordy's & Nicky's Baked Plantain (2)

1 large ripe green or red plantain (sliced in half)

Sauce:
2 parts fresh orange juice
1 part Port/Sauterne (mixed)
1 tsp. honey on each plantain

Grease a glass baking dish. Pour the sauce over the plantain and sprinkle with cinnamon and dot lightly with sweet butter. Bake for 1 hour at 325° to 350°.
13 mg. sodium per half cup
Negligible sodium in plantain

Baked Cauliflower (2)

1 cup sour cream
1 tbl. chopped chives
1 tbl. chopped parsley
1 head of cauliflower
2 tbl. of lemon juice

Boil enough water to cover cauliflower head. Add lemon juice. When water boils add cauliflower and simmer about 20 min. Cooking time depends on the size of the head.
Mix sour cream, chives and parsley. Blend lightly with cooked cauliflower. Place in small glass baking dish and top with bread crumbs. Bake at 350° for ten min. or until slightly browned on top.
36 mg. sodium.

Onions Baked in Foil (4)*

8 medium onions (yellow onions have a stronger taste)
2 tbls. butter
pepper
thyme

Put each onion on a square of tin foil. Sprinkle them with pepper and thyme. Put a lump of butter on each one and wrap the foil around them. Bake at 350° until tender.

12 mg. sodium per serving

Corn Cooked in the Husks

We are accustomed to putting so much salt on our corn! This method of cooking makes that quite unnecessary. Corn cooked this way has a subtle, smokey flavor.

Leave the corn in the husk and bake it in the oven at 400° or under the broiler for about 15 minutes. Serve with unsalted butter and garlic powder (not garlic salt!)

4 mg. sodium per ear

* This recipe also works for mushrooms. You can put a pound of stemless mushrooms together in a foil package with any seasonings you like, and cook them for 15 minutes in a hot oven while you're cooking a steak or roast.

Yogurt Dressing

1 egg yolk
1 cup yogurt
½ tsp. dry mustard
3 tbl. vinegar
½ cup sesame seed oil
pepper to taste (optional)

Combine all ingredients and pour over salad greens.
216 mg. sodium per cup

Mint Dressing

6 tsp. finely chopped mint leaves
1 cup lemon juice
4 tbsp. honey
Combine ingredients in a blender. Great over fruit.
Negligible sodium

Cucumber and Yogurt Dressing (2)

2 cups plain yogurt
1 small onion minced
1 tsp. paprika
1 large cucumber peeled and grated
1 tsp. powdered cumin
½ tsp. ginger powder
Serve over red lentils.
128 mg. sodium per cup

Spaghetti

You can even enjoy spaghetti on a sodium restricted diet. Cook it normally, but of course don't add any salt to the cooking water. 1 cup of cooked spaghetti contains only 3 mg. of sodium. Drain the spaghetti.

Watercress Sauce

1 bunch watercress
½ cup olive oil
6 walnuts
½ teaspoon tarragon
freshly ground pepper to taste

Cut the stems from the watercress and combine all the ingredients in a blender.

24 mg. sodium per cup

French Cream of Watercress Soup

1 bunch watercress
3 tbls. butter
1 bunch leeks
2 medium potatoes, peeled and diced
3 cups water or homemade chicken stock
1 cup milk (light cream optional)
white pepper
nutmeg

Melt the butter in a heavy saucepan. Cut the roots and greens off the leeks and cut them lengthwise and wash them thoroughly. (If you don't do this well, they can be quite gritty and unpleasant in the soup.) Saute the leeks in the butter until they are soft and add the potatoes and stock or water. Simmer this mixture for 20 minutes. Cut the stems off one bunch of watercress and add them to the stock. Cook for another five minutes. Allow the soup to cool and puree it in a blender. Return the puree to the pot and add the cream, pepper and nutmeg. Whisk it until it is smooth and serve. It is delicious hot or cold. You can make the puree a day ahead and refrigerate it overnight. Add the cream just before serving. I think this refines the taste. You may also want to add 2 tbls. of sherry with the cream.

25 mg. sodium per serving.

Mushroom Soup

½ lb. fresh mushrooms
1 cup milk
1 oz. shelled unsalted sunflower seeds
1 clove garlic crushed
3 scallions complete with bulbs.

Combine ingredients in a blender and add water as desired for thickness. Heat and serve.

137 mg. sodium per cup.

Doogh (Cold Persian Soup) (4-6)

Mix together:
1 cup cucumber, chopped and peeled
½ cup minced shallots
¼ cup dried currants (make sure they are not
preserved with sodium sulfite)
Chopped, fresh mint
1 tbl. chopped dill

Add this mixture to:
2 or 3 cups of yogurt
2 tbls. cream

137 mg. sodium per cup

Guacamole (4)

1 large ripe avocado
1 small onion chopped fine
1 clove garlic crushed
1 green pepper chopped
1 tbl. olive oil
1 tbl. wine vinegar
1 tbl. lemon juice
1 tomato (very ripe) peeled and diced

Mash avocado to a semi-lumpy state and add all the other ingredients. To be eaten with crisp crackers or traditional corn chips.

23 mg. sodium per cup.

Chicken

Chicken is still the best buy as far as meat goes so I have given several recipes as well as general ideas about cooking chicken and using left overs. The white meat of the breast is slightly lower in sodium than the dark meat.

Chicken Marinade (2)

½ cup olive oil
1 tbl. vinegar
2 tbls. white wine
¼ tsp. white pepper
½ tsp. rosemary
½ tsp. tarragon
1 clove garlic, crushed
1 bay leaf

Combine all the ingredients and whisk them together well until you have a thick emulsion. Use this marinade on any kind of chicken you like and then broil, bake or roast it. I usually let the chicken sit for 2-6 hours at room temperature.

15 mg. sodium

Herb Butter

Another delicious way to cook chicken is with herb butter. Herb butter can also be added to vegetables or spread on toasted low sodium bread. Herb butter can be made with any herbs you fancy. I store several types in baby food jars in my ice box.

Soften a stick of butter or margarine. Add two tbls. of the herb of your choice, and cream it all together.

Negligible sodium

Broiled Chicken With Herb Butter (2)

Dot two chicken breasts with herb butter and sprinkle. them with pepper. Heat the broiler to high.

Place the chicken under the flame, skin side down. Broil 10 minutes. Turn the chicken, add a little more butter, and broil 15 minutes more. If the chicken is not done, turn it and broil it 5-10 minutes more.

90 mg. sodium per breast

Chicken Stock

Homemade chicken stock really does provide a superior taste to soups and sauces. Gourmets always make their own stocks, and there is no reason why you shouldn't make a salt-free stock. Salt-consuming gourmets in a hurry can always resort to bouillon cubes or canned chicken broth, but these are forbidden to you because they are mainly made of MSG and other sodium compounds. 1 packet of instant chicken broth contains 575 mg. of sodium.

Put the chicken in a large pot (6 to 8 quarts) and enough bold water to cover it by an inch. Throw in an onion stuck with two cloves, 1 carrot, dill, celery, parsley, leeks and whatever herbs you fancy. I recommend a bay leaf and half a teaspoon of thyme. After soup comes to a boil, cover and simmer for at least three hours. Strain out the herbs and onion. You can store the stock for a couple of weeks in the refrigerator or freeze it for later use.

Other Stocks

Stocks may be made in the same way with beef or fish. With fish use leftover skin or heads but not small bones. Fish stock tastes particularly good in vegetable soups and sauces for fish. Beef stock should be used as a base in beef stews and gravies.

Chicken Chasseur (4)

4 chicken breasts
2 tbls. olive oil
2 tbls. minced shallots
¼ cup white wine
½ cup chopped tomatoes
1/8 tsp. white pepper
3/8 cup chicken stock or water
1 bay leaf
1/8 tsp. thyme
1/8 tsp. sweet marjoram
1 cup sliced mushrooms
1 tbl. brandy

Dredge the chicken breasts in flour. Saute it with the shallots until the chicken is golden. Add the other ingredients and simmer covered for an hour. I serve this with new potatoes in their jackets with butter and parsley.

100 mg. sodium per serving

Lemon Veal (4)

1 pound scallopini of veal, pounded thin
½ cup sherry
½ cup flour
2 tbls. butter
1 lemon, sliced
freshly ground pepper

Pound the veal until it is even thinner. Dust it lightly with flour. Melt the butter in the skillet and brown the veal gently. Veal can become tough if the heat is too high or if it is overcooked. Add the sherry and simmer covered for a few minutes until the veal is tender. This should not be more than five minutes. Grind fresh pepper over the veal and garnish it with lemon slices.

142 mg. sodium per serving

Cornish Game Hens (2)

1 cornish hen
3 tbls. butter
½ cup dry white wine
pepper, thyme or rosemary

Rub the hen with one tablespoon of melted butter. Sprinkle it with pepper and thyme or rosemary. Put it in a roasting pan and roast at 350°. Melt the remaining butter and baste the hen with it. When the hen is golden, pour the wine over it and baste it again. Cook until the leg joints move easily. (About one hour.) You may carve the hen or cut it in two.

100 mg. sodium per serving

Marinated Flank Steak (4)

1 lb. flank steak
2 cloves garlic, crushed
3 tbls. honey
½ cup sherry
lemon

Rub the steak on both sides with crushed garlic. Reserve the garlic. Then spread both sides of the steak with honey. Roll it up and put it in a glass or earthen dish. Pour the sherry over it and squeeze a lemon over it. Put the crushed garlic in the marinade, and let it sit at room temperature for 6-12 hours, turning the meat occasionally. Heat the broiler to high. Put the steak under the flame for 4-6 minutes on a side. Make thin, diagonal slices. This meat is also delicious cold.

Roast Leg of Lamb *

This is a simple recipe but it makes the best roast lamb I have ever tasted.
Make about ten slits under the skin of a leg of lamb. Insert a sliver of garlic in each. Rub a lemon over the lamb and sprinkle it generously with rosemary. Roast the lamb at 325° 30-35 minutes per pound.

Three ounces of cooked lamb contains 100 mg. sodium

* Lamb chops are also good rubbed with garlic and sprinkled with either rosemary or thyme. You can broil them or pan fry them in a little butter.

Vesta's Shish-kebab Marinade for Lamb (6)

½ cup olive oil
½ cup Sauterne
1 onion, chopped fine (large size)
1 clove garlic chopped fine
¼ to ½ tsp. pepper
2 tsps. ground cloves
3 large bay leaves
2 tbl. vinegar

Spoon all of the above ingredients over a small leg of lamb (cubed), place in a glass baking dish. Cover and keep in the refrigerator for 48 hours. Turn occasionally. When done stitch on skewers with mushrooms, tomato wedges, green pepper wedge and small white onions in alternating order. Broil in small pan to retain juices, serve over rice using gravy created from broiling.

54 mg. sodium per cup

Lamb Shanks (4)

Lamb (or veal) shanks are delicious and inexpensive. You may have to ask your butcher for them in advance if they are not popular in your area.

4 shanks
1 bay leaf
¼ cup olive oil
½ cup flour
½ cup lemon juice
1 tbl. grated lemon rind
fresh pepper

Roll the shanks in the flour and sprinkle them with pepper. Sear them in the oil for 15 minutes. Then place them in a greased casserole with 1 bay leaf and 4 pepper corns. Add ½ cup of lemon juice to the skillet. Stir to loosen the brown bits on the bottom and pour this over the shanks. Add 1 tbl. grated lemon rind, cover and bake at 350° for 1½ to 2 hours.

Each shank has about 102 mg. sodium

Grilled Fish (2)

2 fillets of any type of fresh fish
(be sure it has not been washed or frozen in brine.)
2 tbls. butter
1 onion
1 lemon
oregano
pepper

Place each fillet on a piece of foil. Dot each piece with 1 tbl. butter. Sprinkle them with pepper. Then put a layer of onion slices and a layer of lemon slices on each. Sprinkle with oregano. Fold the foil over so that each fish is enclosed in a packet of foil. Put the packets under a hot broiler or grill them over a fire for 20 minutes, turning them once.

This is absolutely delicious and can even be cooked on a camping trip.

About 82 mg. sodium per fillet depending on what fish you choose.

Desserts

Fruits are your best bet for desserts, because most of them are very low in sodium and high in potassium. Cakes and cookies often require baking powder which is high in sodium. If you do like to bake, you can get a type of low sodium baking powder at your health food store.

Baked Peaches (4)

4 ripe peaches
1 cup chopped almonds
1 tbl. sherry
½ tbl. lemon juice

Remove the peach skins by dropping the peaches into boiling water for 1 minute and then sliding the skin off. Cut the peaches in half and remove the pits. Place them in a baking dish, cut side up. Moisten the almonds with the sherry and the lemon juice, and stuff this mixture into the peach halves. Bake them for 15-20 minutes at 350°.

9 mg. sodium per peach

Crisp Baked Apples (4)

4 apples
2 tbls. butter
2 tbls. flour
½ cup honey

Peel the apples halfway and core them. Place them in a baking dish. Melt the butter and stir in the flour. Mix thoroughly. Add the honey and spread this mixture over the apples. Sprinkle with cinnamon. Bake them at 400° until the top is crusty, then lower the heat to 325° and bake 35 minutes longer until the apples are tender.

5 mg. sodium per apple

Cameron's Bananas (2)

2 bananas
1 tbl. lemon juice
2 tbls. sugar

Slice the bananas lengthwise. Sprinkle both halves with the lemon juice and spread them with the sugar. Chill them for at least half an hour.

2 mg. sodium per serving

Baby Food Ideas

One of the most urgent aims of KILLER SALT is to ensure that the present and future generations of babies will grow up free from our current national salt addiction. The following recipes are to be used as guidelines in preparing your infant's meals. It is possible to substitute the baby's favorite fruits and vegetables in place of the ones given. None of the recipes contain any added sodium; your baby will never miss it.

Baby Stew Dinner

1 lb. stew meat, sliced into small cubes
½ cup green beans
1 potato, peeled and sliced into small cubes
1 carrot chopped fine
1 cup water
(½ cup peeled and chopped tomatoes if the child is over 9 months.)

Brown the meat in a drop of oil and add all other ingredients. Cook over low heat for an hour adding more water if necessary. You may also cook this in a pressure cooker for 20 minutes.

When the mixture is cooked, add it to the blender in small amounts and puree it.

It is easy to freeze the stew in plastic ice cubes and heat and serve as needed.

90 mg. sodium per ½ cup

Sam Messinger's Baby Borscht

1 cup sliced beets
1 cooked and diced potato
1 cup plain whole milk yogurt

Puree the beets and yogurt in a blender. Mix in the potatoes and serve.

25 mg. sodium per ½ cup

Yogurt Banana Freeze

3 ripe bananas
2 8 oz. containers whole milk yogurt—any flavor

Put the yogurt in the blender. Add the bananas broken into large pieces. Blend. Freeze in a freezer tray and pop a cube out when needed.

Varying flavored yogurts gives variety to the diet and introduces your child to new flavors.

60 mg. sodium per ½ cup

FOODS EXTREMELY HIGH IN SODIUM

Vegetables   *yet salt is added*

Frozen peas
Frozen lima beans
Artichokes
Beet greens
Beets
Celery
Dandelion greens
Kale
Mustard greens
Sauerkraut
Spinach
Chard
White turnips

Fish
Herring
Lobster
Sardines
Shrimp
Scallops
Anchovies
Caviar
Clams
Crabs
Salt fish
Shellfish
Canned fish

Baked Goods
Bread
Cereals
Cakes
Potato chips
Pretzels
Popcorn

Meats
Cold cuts
Brains
Corned Beef
Hot dogs
Kidneys
Sausage
Smoked meats
Canned meats

Miscellaneous
Bouillon cubes
Catsup
Celery salt and flakes
Instant cocoa
Horseradish
Mayonnaise
Meat tenderizers
Molasses
Mustard
Olives
Pickles
Pudding mixes
Junket
Relish
Soy sauce
Worcestershire

SODIUM IN AVERAGE SERVINGS OF BREADS, CEREALS AND FLOURS

Food	*Measure* *	*Sodium Milligrams*
BREAD		
White, enriched, rye, whole wheat	1 slice	138
Unsalted white enriched, rye, whole wheat.	1 slice	7
Unsalted Melba toast, Cellu	1 slice	0.8
CEREALS, COOKED AND DRY		
All-Bran, Kellogg's	½ cup	370
Bran Flakes, £&% Kellogg's	¾ cup	340
Corn Flakes, Kellogg's	1⅓ cups	280
Corn Meal	¾ cup cooked	0.6
Corn Soya, Kellogg's	¾ cup	310
Cream of Wheat, Regular	¾ cup cooked	0.6
Cream of Wheat, Quick	¾ cup cooked	71
Grape-Nuts, Post's	¼ cup	187
Krumbles, Kellogg's	¾ cup	170
Maltex	⅔ cup cooked	1.0
Muffets	1 biscuit	1.0
Muffets	1 biscuit	1.0
Pablum	12 tablespoons	176
Pep, Kellogg's	1 cup	226
Pettijohn's	⅔ cup cooked	0.6
Raisin Bran, Kellogg's	⅔ cup	280
Ralston's Instant and Regular	¾ cup cooked	0.3
Ralston's Rice Chex	1 1/8 cups	230
Ralston's Wheat Chex	½ cup	210
Rice, dry, polished	⅔ cup cooked	0.6
Rice, flakes	1 cup	204
Rice, Krispies, Kellogg's	1 cup	280
Rice, puffed	2 cups	0.3
Rolled Oats	⅔ cup cooked	0.6
Special K, Kellogg's	1¾ cups	193
Sugar Corn Pops, Kellogg's	1 cup	85
Sugar Frosted Flakes, Kellogg's	1 cup	170
Sugar Smacks, Kellogg's	¾ cup	16
Wheatena	⅔ cup cooked	0.6
Wheat flakes	¾ cup	369
Wheat flakes, unsalted	¾ cup	0.6

*1 ounce (28.35 grams) is the weight for dry cereals before cooking.

Wheat germ, Zing .... 1 tablespoon 0.9
Wheat, puffed .... 2 cups 1.0
Wheat, shredded .... 1 large biscuit 0.6
Wheat, whole or cracked .... ¼ cup cooked 0.3

CRACKERS OR WAFERS
   Graham .... 1 cracker 2½" square 25
   Ry-Krisp .... 1 double square 48
   Matzoth, unsalted, plain .... 1 piece, 6" diameter 0.2
   Soda .... 1 cracker 2½" square 66
   Whole Rice Wafers, Cellu .... 2 wafers 0.1

FLOUR
   White enriched .... 1 cup sifted 2.2
   Whole Wheat or rye .... 1 cup stirred 2.2
   Self rising .... 1 cup 1950

OTHER CEREAL PRODUCTS
   Hominy, canned .... ¼ cup cooked 71
   Macaroni and Spaghetti .... ⅔ cup cooked 1.4
   Noodles, egg .... 1 cup cooked 3.0

SODIUM IN AVERAGE SERVINGS OF DAIRY PRODUCTS

Food	Measure	Sodium Milligrams
Buttermilk, cultured	8 ounces	317
CHEESE		
Cheddar or Swiss	1 ounce	198
Cottage	1 ounce	59
Cottage	5 to 6 level talbespoons	290
Cottage, unsalted	5 to 6 level tablespoons	20
Cream, Philadelphia	1 ounce	71
Process	1 ounce	425
Cream	1 ounce	11
Eggs, whole	1 medium	70.2
Ice Cream	1/6 quart	85
MILK		
Fresh, whole or skim	8 ounces	122
Nonfat, dried, reliquified	8 ounces (5 qts. from 1 lb.)	118
Lonalac, reliquified	8 ounces (4 qts. from 1 lb.)	4.3
Walker-Gordon Lo-Sodium	8 ounces	6

SODIUM IN AVERAGE SERVINGS OF
FATS, NUT BUTTER, OILS, AND DRESSINGS

Food	Measure	Sodium Milligrams
Butter, salted	1 tablespoon	140
Butter, unsalted or sweet	1 tablespoon	1.4
Margarine, salted	1 tablespoon	154
Margarine, unsalted	1 tablespoon	1.4
Mayonnaise (average)	1 tablespoon	77
Mayonnaise, no added sodium	1 tablespoon	3.25
Peanut butter	1 tablespoon	18
Peanut butter, no added sodium	1 tablespoon	.75
Oils—corn, cottonseed, olive, cod-liver	1 tablespoon	negligible
Shortening (Crisco, Lard, Spry)	1 tablespoon	negligible
Salad Dressing, French Dietetic	1 tablespoon	2

SODIUM IN AVERAGE SERVINGS OF FRUITS AND FRUIT JUICES

Food	Measure	Sodium Milligrams
Apples, raw	1 small	1
Applesauce, canned	½ cup	2
Apricots, raw	2-3 medium	1
Apricots, unpeeled, canned	4 halves	4
Apricots, dried	4 to 6 halves	3
Banana	1 small	1
Blackberries or Blueberries	5/8 cup	1
Cantaloupe	⅓ of 4½ " melon	13
Cherries, raw, frozen, canned	15 large	2
Dates	3 to 4 pitted	0.3
Figs, raw, canned	2 large or 3 small	2
Figs, dried	2 small	10
Grapefruit, raw	½ small	1
Grapefruit canned in syrup	⅓ cup	2
Grapefruit juice	3 ¼ ounces	2
Grapes, American varieties	1 bunch, 22 to 24	2
Grapes, Thompson seedless	1 bunch, 60	2
Lemon, pulp and juice	1 medium	1
Orange	1 small	1
Orange juice, canned	3 ¼ ounces	1
Peaches, raw	1 medium large	1
Peaches, canned	2 halves, 1 tablespoon juice	2

Peaches, frozen	½ cup, scant	3
Pears, raw and canned	1 medium	2
Pineapple, raw, frozen or canned	½ to ⅔ cup	1
Pineapple juice, canned	3 ¼ ounces	1
Plums, raw or canned	2 medium	1
Prunes, dried	6 medium	3
Prune juice	3 ¼ ounces	2
Raisins	1 tablespoon	2.5
Raspberries, raw	⅔ to ¾ cup	1
Raspberries, canned	½ cup scant	2
Rhubarb, raw or frozen	¾ to 1 cup cubes	2
Strawberries, raw	10 large	1
Strawberries, frozen	½ cup scant	2
Tangerines	1 large or 2 small	1
Watermelon	½ cup cubes or balls	1

SODIUM IN AVERAGE SERVINGS OF MEATS, FISH AND POULTRY

Food	Measure	Sodium Milligrams
Bacon, fried crisp	3 strips	552
Beef, lean muscle, raw	3½ ounces	70
Chicken (average light and dark meat)	3½ ounces	75
Catfish. raw, unsalted	3½ ounces	60
Codfish, raw, unsalted	3½ ounces	65
Codfish, dried, salted	3½ ounces	8100
Crab, boiled	3½ ounces	370
Duck	3½ ounces	85
Halibut, raw, unsalted	3½ ounces	56
Ham, cured	3½ ounces	1100
Frankfurter, cooked	1 average	550
Lamb, lean, raw	3½ ounces	90
Liver, beef, raw	3½ ounces	130
Liver, pork, raw	3½ ounces	80
Liver, calf, raw	3½ ounces	110
Lobster, boiled	3½ ounces	250
Pork, lean, raw	3½ ounces	55
Salmon, canned	3½ ounces	540
Salmon, canned without added sodium	3½ ounces	60
Sardines. canned	3½ ounces	550
Sausage, bologna	2 slices	780
Sausage, pork	3½ ounces	750
Shrimp, raw	3½ ounces	140
Tuna, canned, drained	5/8 cup	800

Tuna, canned without salt	5/8 cup	50
Turkey, raw...................................	3½ ounces...............	65
Veal, raw.....................................	3½ ounces...............	100

SODIUM IN AVERAGE SERVINGS OF VEGETABLES

Food	Measure	Sodium Milligrams
Asparagus, raw	5 to 6 stalks	3
Asparagus, canned, spears	6 medium stalks	410
Asparagus, canned without added sodium ...	6 medium stalks	4
Asparagus, frozen	6 medium stalks	13
Beans, dry, navy or pea	¼ cup	0.5
Beans, lima, raw	4 rounded tablespoons ...	1
Beans, lima, canned, drained solids	½ cup	248
Beans, lima, canned without added sodium ...	½ cup	1.6
Beans, snap, green and yellow wax, raw	1 cup	1
Beans, snap, canned	½ cup	258
Beans, snap, canned without added sodium ..	½ cup	1.2
Beans, snap, frozen	½ cup	2.0
Beets, raw....................................	2 beets 2″ diameter	60
Beets, canned	½ cup	30
Beets, canned without added sodium	½ cup	33
Beet greens, raw	½ cup cooked.............	130
Broccoli, raw and frozen	1 stalk 5½″ long	15
Brussels Sprouts, raw and frozen	9 medium.................	12
Cabbage, raw	½ cup shredded	7.5
Carrots, raw	1 large or 2 small	50
Carrots, canned without added sodium		
Carrots, canned wtihout added sodium	½ cup	26
Cauliflower, raw and frozen	½ cup cooked.............	12
Celery	3 small stalks; ½ cup diced................	50
Chard, raw	⅓ to ½ cup cooked	100
Corn, sweet, raw	1 medium ear.............	1
Corn, sweet, canned..........................	½ cup	166
Corn, sweet, canned without added sodium.....................................	½ cup	1.7
Cucumber, raw	6 to 8 slices - ½ medium ..	2.5
Kale	1 cup scant cooked	80
Lettuce, head	4 small or 2 large leaves ..	7.5
Mushrooms, raw	10 small, 4 large	5
Mushrooms, canned	½ cup solids and liquids..	488
Okra, raw	8 to 9 pods	1

Olives, pickled, green	1 large	130
Onions	1 onion 2¼″ diameter	10
Peas, raw, canned without added salt	½ cup cooked	1.6
Peas, canned, drained solids	½ cup	216
Peas, frozen	½ cup variable	
Peas, dried	¼ cup	10
Peppers, green, raw	1 medium	0.8
Potatoes, white, raw	1 potato 2¼″ diameter	3
Radish	1 small, 1″ diameter	1.5
Sauerkraut, canned	⅔ cup	650
Spinach, raw	½ cup scant, cooked	75
Spinach, canned without added sodium	½ cup	45
Spinach, frozen	½ cup scant, cooked	70
Squash, raw, all types	½ cup	1
Sweet potato, raw	1 small	10
Tomato, raw	1 small	3
Tomato, canned	½ cup, scant	18
Tomato, canned without added sodium	½ cup, scant	3
Tomato juice, canned	⅓ to ½ cup	230
Tomato juice, canned without added sodium	⅓ to ½ cup	5
Turnips, white	⅔ cup cooked	40
Turnips, yellow	⅔ cup cooked	5
Turnip greens	½ cup cooked	10

SODIUM IN AVERAGE SERVINGS OF MISCELLANEOUS ITEMS

Food	Measure	Sodium Milligrams
Baking powder, ordinary	1 teaspoon	450
Baking powder, low sodium	1 teaspoon	2
Cocoa, plain, Hershey	1 ounce or 4 tablespoons	1
Cocoa, Dutch process	1 ounce or 4 tablespoons	17
Chocolate, Bitter	1 ounce	3
Coconut, raw	¼ cup shredded	7
Coconut, shredded, dried	¼ cup	3
Coffee, roasted dry	1 ounce	0.6
Gelatine, plain, dry	1 tablespoon	3
Gelatine dessert, flavored	1 box	280
Tapioca, dry	1 tablespoon	0.4
Tea, blend, dry	1 ounce	1
Vinegar	1 tablespoon	negligible
Yeast, bakers	1 cake	0.4

CHAPTER ELEVEN

Important Questions & Answers About Salt

Q. *What Is Salt?*

What we call salt is actually sodium chloride, a compound of sodium and chlorine. Salt and sodium are sometimes used interchangeably. An excess of any sodium compound can be damaging. Other sodium compounds include monosodium glutamate and sodium nitrite. Iodized salt is table salt with iodine added to it in order to prevent goiter.

Q. *Is There A Daily Requirement For Salt?*

The daily requirement for sodium has not been established because it occurs so plentifully in so many foods. Certainly 500–1,000 mg. of salt per day is enough to maintain the correct sodium balance in your body. This amount is equal to ¼-½ teaspoon daily. If you eat meat or dairy products you are adequately supplied with sodium and need never use table salt again. If in doubt, ask your doctor how much you need on the basis of your age and medical history.

Q. *What Foods Provide Sodium?*

Sodium is most plentiful in meat, fish, and dairy products. It is also plentiful, unfortunately, in processed foods. It is important to eat fresh food for your sodium requirement, because you get the benefit of the minerals and vitamins that are usually washed out during processing, and often the sodium compounds in processed food add too much sodium to your diet.

Q. *How Do I Know If I Am Getting Too Much Salt?*

There may be a rise in your blood pressure which will show up when your doctor examines you. If you notice rolls of tight, water-filled skin around your knuckles and ankles, you are probably retaining water because there is too much sodium in your diet, or because for some reason your body is retaining all the sodium you consume. This fluid spillover into the tissues is called edema.

Q. *How Do I Know If I Am Not Getting Enough Salt?*

First of all, it is hard to imagine anyone living on the typical American diet of meats and dairy products not getting enough salt. There is enough sodium in meat and dairy products to meet your daily requirements for salt as long as you are eating enough of them to nourish yourself. The adult requirement for sodium has not been established, because sodium is so plentiful in the modern western diet. A sodium deficiency would be apparent through extreme muscular weakness and cramping. The symptoms could include nausea or diarrhea. Sodium deficiency can occur during adrenal failure, because the kidneys are not conserving any sodium and the body's stores are excreted in the urine. This condition must be treated promptly or death can ensue. Fortunately, it is very rare.

Q. *If We Only Need A Little Salt*
  Why Does Everyone Use It So Much?

Most people use salt out of habit which was probably instilled in them from childhood. Perhaps it is also a carryover from the early days of mankind when salt was a valued commodity and no one had as much of it as he wanted. In time it began to be a symbol of wealth. So much of the food we eat is so bland and its preparation so unimaginative that the only way some people know to rescue the flavor and make it palatable is by adding salt to it. In the recipe section of this book you will find a chart of herbs which may provide some inspiration so that salt will no longer be the seasoning you most use.

Q. *How Much Salt Does a Baby Need?*

A baby needs a very, very small amount of salt per day. It is very difficult to make a baby deficient in sodium, because mother's milk contains a sufficient amount and cow's milk contains more than enough. Rather than worrying about providing enough salt for your baby, try instead not to add any more to his diet. Keep away from commercial baby foods and prepare your own from fresh vegetables and meats. As the child grows, do not introduce him to table salt or sodium-rich foods which do not provide much nutrition such as bread and rolls. Not introducing bread into the diet will also keep your child thinner, because he will be cutting down on one of the major sources of starch in the diet. More and more doctors are reporting that a child will be healthier if he is slightly on the thin side and acquires good eating habits.

Q. *Does Salt Have Something To Do With The Anxiety And Depression That I Feel Just Before My Period?*

Yes, it does. Some women have a tendency to retain almost all of the sodium they eat instead of excreting it normally in their urine. This retention of sodium leads to a build-up of fluids. These fluids press on all the tissues in the body, even those of the brain, and can cause extreme irritability and depression. The symptoms of the disorder vary from woman to woman—some women even become mildly psychotic during the premenstrual week, although this is rare. Mild headaches and cramps are common physical symptoms during this time. It is not fully understood what causes some women to retain fluids and others not to, but doctors are beginning to believe that it is related to the balance of female hormones—a delicate matter since the system of balance is so complex among the various hormones.

Q. *What Can I Do If I Suffer From Premenstrual Tension?*

Try restricting your intake of sodium and fluids during the period when you usually first experience symptoms of distress. If you are eating meat and dairy products, you are taking in sufficient sodium for your body's needs, and you don't have to add another shake during cooking or at the table. This may disagree with your taste buds at first, but I am sure your symptoms will decrease if not disappear entirely. If you still have annoying symptoms, go to your doctor; he may want to prescribe a diuretic.

Q. *What Are Diuretics?*

Diuretics are a group of medications which cause your kidneys to excrete water. There are several different types of diuretics and which one your doctor will prescribe depends upon the condition you are being treated for. Some diuretics cause you to lose potassium, and you may need to compensate for this loss in some way. *Never take diuretics except under your physician's advice.* The kind sold over the counter usually contain ammonium chloride, which washes sodium from your body. However, nutrients are washed out along with the salt, and the body can suffer seriously from this.

Q. *I've Heard That Diuretics Help You Lose Weight.*

This is not really true. Diuretics wash out salt and cause fluids to be excreted. This will result in an apparent weight loss of two to five pounds immediately, but you are not losing an ounce of fat this way. Some diuretics work directly on the kidneys, an important organ that you should not be playing with. Diuretics are not satisfactory for long-term use during a diet, because the pills rapidly lose effectiveness in the body.

Q. *Should I Eat Less Salt If I Am On The Pill?*

Yes. A major side effect of the Pill seems to be hypertension or an increase in the blood pressure even to the high normal range. Pill-induced hypertension seems to be a result of increased blood volume due to the retention of sodium and water. Progesterone, a female hormone and one of the major ingredients of the Pill, can cause the body to retain sodium and fluids. Little is known about this yet, but women on the Pill are more likely than others to develop elevated blood pressure, and anyone for whom this risk exists, either because of a familial tendency or because they are on the Pill, would do well to limit their salt intake and learn to enjoy food without salt before they develop any disorders.

Q. *What Does Salt Have To Do With High*
   Blood Pressure?

Doctors believe that blood pressure can increase in two ways: by constriction of the blood vessels, or by an increase in the volume of the blood. It is this second type of high blood pressure which is more closely related to salt. Salt causes water to be retained instead of being excreted normally by the kidneys. This retained water increases the amount of blood in circulation, thus driving up the pressure. The heart has to work harder to pump the increased amount of blood through arteries which have not expanded and may have contracted due to the presence of other chemicals. Some animal studies have shown that a high salt diet in itself is enough to bring on high blood pressure in experimental rats which are genetically predisposed to hypertension. The analogy here is to persons who unknowingly are predisposed to hypertension because of heredity. For such people, a high salt diet can be a trigger at a very early age for problems to begin.

Q. *What Can I Do About Hypertension?*

Nothing on your own except cut down the amount of salt you eat. It is essential to have a checkup regularly, because hypertension is sometimes asymptomatic or its symptoms, if they exist, are minor annoyances such as headaches. If you are diagnosed as having hypertension, you must stay under a doctor's supervision. There are so many types of hypertension that self-diagnosis is foolhardy and dangerous, because you will probably need medication as well as a low sodium diet. It is crazy not to take ten minutes once or twice a year to have your blood pressure checked. It might save your life. Cutting dietary salt drastically can be a good preventive measure.

Q. *Why Do They Add Salt to Canned Food?*

Mainly to cover up the bland, overprocessed taste. In some cases a product cannot be preserved by treating it at high temperatures; in this case a sodium preservative is added.

Q. *I Am On A Low Salt Diet. Why Can't I Eat Canned Soup and Vegetables?*

You may if you can calculate the exact amount of sodium the can contains and subtract it from your daily sodium allowance. However, these products are very heavily salted. This is true of vegetable juices as well. One half cup of tomato juice contains 200 mg. of sodium. Canned soups not only contain a great deal of sodium chloride, but also contain MSG or another of the sodium compounds. Sodium is needed in these products to spark up the dreadfully bland taste. After you have learned to cook fresh food for yourself and spice it creatively, you will never want to eat canned vegetables or soups again.

Q. *Is It All Right To Eat Frozen Fruits And Vegetables?*

Frozen fruits are usually free of sodium unless they have a sauce which may contain a sodium compound. Frozen vegetables are generally safe to eat unless they are in a butter sauce. Peas and lima beans often have salt added. *Read the label!*

Q. *Is It All Right To Use Salt Substitutes?*

Only on your doctor's recommendation. Some of them contain potassium. It is not a good idea to start meddling with the body's sodium-potassium balance, particularly if you are having problems with salt. Some salt substitutes actually contain sodium in small quantities, so use them only on your physician's advice and ask him for a brand name.

Q. *I've Been Told That Pork Products Such As Bacon*
   Can Be Dangerous. Why?

There are two reasons: 1. These foods are very high in sodium, so they pose an obvious threat to a person who must restrict his salt intake. 2. The sodium compound used in preserving meats such as bacon, ham, and sausage and to give the product a "cured" taste, is sodium nitrite. Sodium nitrite is used to prevent meat from spoiling, to make it redder and to give it the distinctive cured taste that so many people expect from this type of meat. The danger of sodium nitrite is that it combines with certain chemicals found in many foods to form nitrosamines, a group of the most potent carcinogens known to man.

Although sodium nitrite is allowed in meat, do not believe that whatever is good for the meat industry is naturally good for you. Stay away from products containing sodium nitrite as much as possible, because it accumulates in your body and seems to be even more effective at producing tumors when it is consumed in small doses over a long period of time than if given in one large dose.

Q. *Are There Any Medical Uses For Salt?*

Salt has always been used as a medicine though sometimes in ways we might find strange now that we have so many more specific medications. The disinfectant value of salt has always been known. For this reason, salt water makes an excellent gargle for a sore throat. Topically applied salt creates a sort of heat and increases blood flow to a particular area. Sea water is often beneficial for sore muscles, and often injured racehorses are allowed to stand in it. Some dermatologists find that a special preparation of salt is helpful in drying up acne.

CHAPTER TWELVE

A Brief Glossary Of Terms

ACTH: A hormone secreted by the anterior lobe of the pituitary gland. ACTH in turn activates the adrenal cortex to produce steroids.

ADRENAL CORTEX: The outer wrapping of the adrenal gland. This part of the gland is responsible for the production of aldosterone and cortisone-like substances which aid the body in repairing itself.

ADRENAL GLAND: The adrenals are small glands located directly above the kidneys. The work in conjunction with the pituitary to prepare the body to meet the demands of a particular stressor.

ADRENAL MEDULLA: The inner portion of the adrenal gland which secretes norepinephrine (NE) and epinephrine.

ALDOSTERONE: A hormone secreted by the adrenal cortex. It is one of a group of hormones known as mineralacorticoids because of its action on minerals such as salt and potassium. It is released during stress in response to messages from the pituitary-adrenal system. Its release is also triggered by the kidneys when the blood pressure falls too low.

ANGIOTENSIN: A group of proteins thought to be present in the blood and activated by renin. Angiotensin II is thought to raise blood pressure by constricting the walls of the blood vessels as well as by causing fluids to be retained in the blood.

ARTERIOSCLEROSIS: A condition which involves a hardening and a decrease in elasticity of the arteries.

ARTERIOLES: Smaller branches of the arteries. They are invisible to the unaided eye.

ARTERY: A large blood vessel which carries oxygen-rich

blood away from the heart to the tissues of the body.

ATHEROSCLEROSIS: Scarring of the larger arteries which often also become plugged with cholesterol and fatty cells.

BLOOD PRESSURE: The pressure with which the blood is forced through the blood vessels.

CAPILLARIES: Smaller blood vessels than the arterioles. Capillaries connect the arteries with the venules which carry blood back to the heart. Nutrients and waste matter pass through the semipermeable membranes of the capillaries.

CARCINOGENIC: Cancer producing.

CARDIOVASCULAR: Pertaining to the heart and blood vessels.

CORTISONE: An anti-inflammatory drug.

DEGENERATIVE DISEASE: A disease which causes degeneration or deterioration of a certain part of the body. A degenerative disease can occur in any organ of the body.

DIASTOLIC PRESSURE: The resting blood pressure when the heart is not expelling blood. The pressure between pulses of the heart. The pressure within the arterial system is at its lowest point during the diastole.

ECLAMPSIA: A dysfunction of pregnancy characterized by extremely high blood pressure, convulsions and sometimes coma and death.

EDEMA: A condition which occurs when excess fluid leaks into the tissues of the body and causes a visible swelling.

ENDOCRINE: Pertaining to glands which secrete their hormones into the blood which transports them to their target organ. The pituitary and adrenal glands are both endocrine glands.

EPINEPHRINE: An adrenal hormone which acts to prepare the body for emergencies.

ESSENTIAL HYPERTENSION: Hypertension occurring without discernible organic cause.

ESTROGEN: A female hormone which controls the development of female sex characteristics. Estrogen prepares the body for pregnancy and nursing.

GENERAL ADAPTATION SYNDROME: A syndrome first described by Dr. Hans Selye, an Austrian working in Canada. According to his definition; GAS is "The manifestations of stress in the whole body as the develop in time." The GAS evolves in three stages: The stage of alarm, the stage of resistance and the stage of exhaustion.

GOITER: An enlargement of the thyroid gland resulting from an iodine deficiency.

GLOMERULI: The tiny balls of capillaries which cover the outer portion of the kidney.

HORMONE: A chemical substance produced by a specific gland which acts upon a specific target area of the body. Hormones are transported by the blood from their organ of origin to their target.

HYPERTENSION: The correct term for high blood pressure. A condition of abnormally high pressure within the cardiovascular system.

HYPONATREMIA: A condition where there is too little salt in the blood.

HYPOTHALAMUS: A part of the brain which controls water balance, sleep, hunger, and other basic functions. Another part of the hypothalamus sends hormonal messages to the pituitary and mediates deep-seated emotions.

KIDNEY: A double-lobed gland in the area of the lower back which secretes urine and controls the excretory functions of the body. The kidneys can also secrete substances called *renal pressor substances* which are capable of raising the blood pressure.

LABILE HYPERTENSION: A condition in which the blood

pressure rises when a person is under stress or is angry, but falls back to the normal range when the person is resting or not under pressure.

LITHIUM: A mineral once used as a salt substitute. Now used in the regulation of manic-depressive disorders.

MALIGNANT HYPERTENSION: Severe hypertension involving organs such as the kidneys or brain. Malignant hypertension has a very poor prognosis unless treated.

METABOLISM: The chemical and physical processes by which the body converts food to energy.

MONOSODIUM GLUTAMATE (MSG): A sodium compound used in many foods, particularly in Chinese cooking as a "flavor enhancer."

NEPHRITIS: Inflammation of the kidneys.

NITROSAMINES: A group of carcinogenic chemicals which are formed from sodium nitrite, a meat preservative.

NOREPINEPHRINE (NE): A chemical which is secreted by the adrenal medulla in response to low blood pressure. NE also occurs in certain nerve endings in the brain and acts to transmit nerve impulses from one nerve ending to the next.

PITUITARY GLAND: A gland whose hormones control the activities of the adrenal cortex. Another part of the pituitary controls the secretion of growth hormones.

POTASSIUM: A soft metal which acts in the body in conjunction with sodium to regulate the conduction of nerve impulses and the flow of nutrients and waste products in and out of cells.

PROGESTERONE: A female hormone which prepares the lining of the uterus to receive a fertilized egg. Progesterone also makes nutrients available to the developing egg by regulating the storage of fat and proteins.

RENAL HYPERTENSION: Hypertension resulting secondary to any sort of kidney damage.

RENIN: A renal pressor substance (a hormone secreted by

the kidneys in response to a drop in the blood pressure).

SODIUM CHLORIDE: The white crystal which we call salt.

SODIUM NITRITE: A preservative for meat which, in combination with other naturally occurring substances, can cause the production of tumors.

SPHYGMOMANOMETER: The instrument your doctor uses to measure your blood pressure.

STRESS: Dr. Selye's description: "The bodily changes produced whether a person is exposed to nervous tension, physical injury, infection, cold, heat, x-rays or anything else are what we call stress."

SYSTOLIC PRESSURE: The pressure with which the blood is pumped out of the heart.

THYROID: A gland located in the neck which controls the rate of metabolism in the body.

BIBLIOGRAPHY

Bagg, Elma W. *Cooking Without a Grain of Salt.* Doubleday, Garden City. 1964

Beaton, George H. and McHenry, Earle Williard. *Nutrition: A Comprehensive Textbook.* Volume II. Academic Press, New York. 1964.

Birnbaum, Stanley. "Fluid and Electrolyte Metabolism in the Gynecologic Patient" *Clinical Obstetrics and Gynecology.* Vol. 7. No. 1. pp. 178-183

Bowes, Anna dePlanter and Church, Charles. *Food Values of Portions Commonly Used.* PP. 102. College Offset Press. Philadelphia. 1956

Bunge, G. in *The Chemistry of Food and Nutrition.* Ed. A.C. Sherman. pp. 243. Macmillan, New York. 1933

Burton, Benjamen T. "Current Concepts of Nutrition and Diet in Diseases of the Kidney."*J. of the American Dietetic Association.* Vol. 65. Dec. 1974 pp. 627

Consumers Reports July 1973. Vol. 38. pp. 460-4

Dahl, L.K. and Love, R.A. "Etiologic Role of Sodium Chloride Intake in Essential Hypertension in Humans." *JAMA* Vol. 164 pp. 397 (1957)

Dahl, Lewis K. *Journal of Experimental Medicine.* Vol. 114 (1961). pp. 231

Dahl, Lewis K. *Ibid* vol. 115. pp. 1173 (1962)

Dahl, L.K. and Heine, M. and Tassinari, L. "Effects of Chronic Excess Salt Ingestion: Role of Genetic Factors in Both DOCA—Salt and Renal Hypertension." *Journal of Experimental Medicine.* Vol. 118: pp. 605. (1963)

Dahl, L.K. et al. *Nature.* Vol. 198. pp. 1204-5. (1963)

Dahl, Lewis K. et al. *Journal of Experimental Medicine.* Vol. 122 pp. 535 (1965)

Dahl, Lewis K. *Nutrition Reviews.* Vol. 25, pp. 82. (1967)

Dahl, Lewis K. *Ibid* Vol. 26 #4. April 1968.

DeLaney, Janice, Lupton, Mary Jane and Toth, Emily. *The Curse*. E.P. Dutton & Co. New York, N.Y. 1976

Diamond, Sharon B. et al. "Menstrual Problems in Women with Primary Affective Illness." *Comprehensive Psychiatry*. Vol. 17 number 4. July/August 1976

Duncan, Garfield. *Science News Letter*. June 29, 1963

Eliade, Mircea. *Rites and Symbols of Initiation*. Harper 1968.

Fieve, R.R. "Clinical Controversies and Theoretical Mode of Action of Lithium Carbonate." *International Pharmacopsychiatry*. Vol. 5. pp. 107-118. (1970)

Fieve, R.R. *Moodswing: The Third Revolution in Psychiatry*. William Morrow. 1975

Freis, Edward D. "Salt, Volume, and the Prevention of Hypertension." Circulation Vol. 53 Number 4. April 1976

Gardner, Hugh. "Sowbelly Blues. The Links Between Bacon and Cancer." *Esquire*. November 1976. pp. 112

Hanssen, Maurice. *Everything You Want To Know About Salt*. Pyramid Books. New York 1968

Helmer, O.H. *JAMA*. Vol. 204 April 29, 1968.

Holmes, T.H. and Rahe, R.H. "The Social Readjustment Rating Scale." *Journal of Psychosomatic Research*. Vol. 11 pp. 213. (1967)

Journal of the American Dietetic Association. Abstract. Vol. 66. pp. 192. Feb. 1975

Kraske, Robert. *Crystals of Life*. Doubleday. 1968

Laragh, John. "Hormone Profiles and the Treatment of Hypertensive Diseases." *Practical Management of Hypertension*. pp. 27-40.

Laragh, John H. "The Renin Axis in Causation, Prog-

nosis and Treatment of Hypertensive Diseases."
Transactions and Studies of the College of Physicians of Philadelphia. Vol. 41 number 2. October, 1973.

Laragh, John. "Contraceptive Hypertension From 1967 On." *Hospital Practice.* May, 1975.

Meneely, George R. and Tucker, Robert et al. *Annals of Internal Medicine.* Vol. 39 Number 5 pp. 991-8 November 1953.

Meneely, Tucker et al. *Journal of Experimental Medicine* Vol. 98 p. 71. (1953)

Newsweek "Salt Compulsives." Vol. 82. p. 76. Sept. 24, 1973

Newsweek p. 63 June 17, 1974.

Nutrition Reviews Vol. 11. February 1953. p. 33-35

Page, Irvine H. *Hypertension.* Charles C. Thomas. Springfield, Ill. 1956.

Page, L.B. et al. "Antecedants of Cardiovascular Disease in Six Solomon Island Societies." *Circulation.* Vol. 49 p. 1132. (1974)

"The Physician and the Community" A Report of the Hartford Foundation Conference on Ambulatory Care and Rehabilitation." 1964.

Pike, Ruth L. and Brown, Myrtle. *Nutrition: An Integrated Approach.* P. 79. John Wiley and Sons. New York. 1975.

Pitts. *Physiology of the Kidney and Body Fluids.* Yearbook. Chicago. 1963.

Platman, S.R., and Fieve, R.R. "Lithium Retention and Excretion." *Archives of General Psychiatry.* Vol. 20. March 1969.

Prior, I.A.M. and Evans, J.G. "Sodium Intake and Blood Pressure in Pacific Populations." *Israeli Journal of Medical Science.* Vol. 5. p. 608. (1969)

Sasaki, N. *Geriatrics.* Vol. 19. p. 735. October 1964.

Schecter, Paul J., Horwitz, David and Henkin, Robert I. "Sodium Chloride Preference in Essential Hypertension." *JAMA*. Vol. 225: 1311-5. Sep. 10, 1973.

Selye, Hans. *The Stress of Life*. McGraw Hill. New York 1956

Shaper, A.G. et al. "Environmental Effects on the Body Build, Blood Pressure and Blood Chemistry of Nomadic Warriors Serving in the Army in Kenya." *East African Medical Journal*. Vol. 46 p. 282 (1969)

Shubsachs, S. *Lancet*. p. 109. January 9, 1965

Smith, John R. *Nutrition Reviews*. Vol. 11. p. 33 (1953)

Snively, W.D. *JAMA*. April 29, 1968. P. 343.

Tobian, Louis. "Hypertension and the Kidney." *Archives of Internal Medicine*. Vol. 133. p. 959. June 1974.

Weidegger, Paula. *Menstruation and Menopause*.

Weinsier, Roland L. *Preventive Medicine*. Vol. 5 p. 7-14. March 1976.

Weiss, Jay M. "Psychological Factors in Stress and Disease" *Scientific American*. Vol. 226. June 1974

Wotman, Stephen, Mandel, Irwin D., Thompson, Robert H. and Laragh, John L. "Salivary Electrolytes and Salt Taste Thresholds in Hypertension." *Journal of Chronic Diseases*. Vol. 20. p. 833-840. (1967)